Nonlinear Acoustical Imaging

Woon Siong Gan

Nonlinear Acoustical Imaging

 Springer

Woon Siong Gan
Acoustical Technologies Singapore Pte Ltd.
Singapore, Singapore

ISBN 978-981-16-7014-5   ISBN 978-981-16-7015-2   (eBook)
https://doi.org/10.1007/978-981-16-7015-2

This Springer imprint is published by the registered company Springer Nature Singapore Pte Ltd.
The registered company address is: 152 Beach Road, #21-01/04 Gateway East, Singapore 189721, Singapore

To my parents

Foreword

Nonlinearity in acoustics is usually something that is to be avoided. For instance, the nonlinear distortion in a music reproduction system indicates a possible damage in may be the amplifier or loudspeaker. This distortion appears disturbing to us, but it does also contain information about the structural inhomogeneities—like cracks.

This is one example of when a nonlinearity in acoustics yields information about a material or medium. It also introduces the importance in that a small inhomogeneity may result in a heavily nonlinear distortion. Nonlinear acoustic imaging does not follow the same rules and guidelines as normal linear acoustic imaging. For example, material disturbances which are orders of magnitude smaller in size than the acoustic wavelength do not have an impact in the linear techniques, but can give clear indications using nonlinear acoustic methods.

This is why Woon Siong Gan is treating the subject of *nonlinear* acoustic imaging in this specifically dedicated book, despite that he has given an extensive overview of the linear aspects in the previously published book *Acoustical Imaging: Techniques & Applications* (2012).

This book presented the equations and basics of the nonlinear acoustics, its pertinent phenomena and how these are used in imaging techniques for different applications.

The subject of nonlinear acoustic imaging techniques is relatively new and has not been extensively covered in previous books. It is a difficult science to delve into, especially while most acoustic courses teach only the linear acoustics. This is perhaps completely natural, as the move into nonlinear studies is not always quick and easy since some of the accepted conceptions need to be elaborated.

Woon Siong Gan has, during the last years, been prolific in sharing a fraction of his deep and diverse knowledge through publishing the following books: *New Acoustics Based on Metamaterials* (2018); *Gauge Invariance Approach to Acoustic Fields* (2019); *Signal Processing and Image Processing for Acoustical Imaging* (2020); *Time Reversal Acoustics* (2021).

The titles and the contents of all these books show the nature of Woon Siong Gan through how he connects the topical broad spectrum all the way from the most fundamental science to useful application practices—as he has done also in this his latest very valuable book.

Claes Hedberg
Professor, Blekinge Institute
of Technology
Karlskrona, Sweden

Preface

To study nonlinear acoustical imaging, one needs to understand in depth the physics of nonlinear acoustics. Hence, it is necessary to start with the theoretical foundation of nonlinear acoustics. The most popular equation of nonlinear acoustics, the KZK equation only accounts for diffraction, nonlinearity and absorption in directional sound beams. There are several other aspects of nonlinear acoustics which are not covered by this equation. An example is the curvilinear path of high intensity acoustic wavefields. To account for the curvilinear path, the curvilinear spacetime coordinates have to be used. The most well-known application of curvilinear spacetime is the general theory of relativity by Albert Einstein to account for the curvilinear path of the nonlinear gravitational field. The curvilinear spacetime approach starts with the intrinsic nonlinear nature of the problem instead of extending the linear case to nonlinearity by adding on higher-order terms which is still an approximation.

Two well-known unsolved problems of nonlinear acoustics, turbulence and sonoluminescence are phase transition in nature. To treat phase transition problem, statistical mechanics has to be used to study the critical point of phase transition and the singularity behaviour of the transport properties of the region surrounding the critical point. So far the most acceptable theories of turbulence are that due to Kolmogorov which is a statistical theory and statistical mechanics has to be used. The other theory is that of the description of turbulence as a critical phenomenon, a form of phase transition and the solving of the singularity problem of the correlation length at the critical point by the use of renormalization group method which is used in statistical mechanics.

Gauge theory is useful for solving several problems in nonlinear acoustics such as multiple scattering, diffraction and electron–phonon interaction. The symmetry used in gauge theory can simplify several complexities in nonlinear acoustics. Also the usual method used for electron–phonon interaction is many body theory of Dirac which can handle only up to the second-order term. Beyond that, divergence occurs, and there are infinities terms. With the use of gauge theory, the renormalization method can cancel off the infinities terms.

During the last forty years, several formats of nonlinear acoustical imaging have been developed such as harmonics imaging, fractal imaging, B/A nonlinear parameter acoustical imaging, non-classical nonlinear acoustical imaging and modulation method in nonlinear acoustical imaging. The advantages of nonlinear acoustical imaging are high sensitivity and higher image resolution. They can be applied to non-destructive testing, medical ultrasound and underwater acoustics.

Singapore   Woon Siong Gan
September 2021

Contents

Chapter 1
Introduction to Nonlinear Acoustics

1.1 Introduction

Nonlinear acoustics is a branch of acoustics dealing with sound waves of sufficiently large amplitudes. This study will require the full form of the governing equations of fluid dynamics for sound wave propagation in liquids and gases and elasticity for sound wave propagation in solids. These equations are generally nonlinear and linearization is no longer possible for dealing with large amplitude sound waves. The solutions of these equations also show that sound waves are being distorted as they propagate due to the effect of nonlinearity.

In general, the world is of nonlinear in nature. For sound propagation in fluids (liquids and gases), the full nonlinear equation of fluid mechanics will be involved. For propagation in solids, the full equation of elasticity will be used. Practical examples of nonlinear acoustics are shock wave, cavitation, high intensity focused ultrasound (HIFU) and music. Ultrasonic waves commonly display nonlinear propagation behaviour due to their relatively high amplitude to wavelength ratio.

The nonlinear behaviour of sound wave is due to both the nonlinear nature of the propagating sound wave and also the nonlinear nature of the medium of propagation. The nonlinear nature of sound wave means that the propagating sound wave must have large amplitude and the nonlinearity of the medium of propagation means medium also to generate harmonics such as in medical ultrasound imaging [1] due to the nonlinear nature of the human tissue. This means nonlinearity will be generated even with an ordinary intensity sound and without the necessity of a high intensity sound wave. It has also been known that cracks in metals also generate harmonics with ordinary intensity sound wave [2]. The nonlinear nature of sound wave itself and the nonlinear nature of the propagating medium situation is analogous to that of symmetry can be due to the symmetry property of the sound wave and can also be due to the symmetry nature of the medium of propagation for instance crystals.

The nonlinear nature of the medium can be described by designating a nonlinear parameter to describe the medium such as the B/A nonlinear parameter. Sofar B/A

© Springer Nature Singapore Pte Ltd. 2021
W. S. Gan, *Nonlinear Acoustical Imaging*,
https://doi.org/10.1007/978-981-16-7015-2_1

nonlinear parameter has been used mostly for human tissues. In this chapter we extend its use to metals.

1.2   Constitutive Equations

The constitutive equations of nonlinear acoustics can be divided into two categories: one is for sound propagation in fluids (liquids and gases) and the other is for sound propagation in solids. For sound propagation in fluids, the constitutive equation is based on the Navier Stokes equations. The popular equations of nonlinear acoustics in fluids are the Westervelt equation, the Burgers' equation, and the KZK equation. They can be derived from the basic equations of fluid mechanics. For sound propagation in solids, the nonlinear elasticity equation has to be used. All these equations are analysed in more details in the subsequent chapters of this book.

Harmonics generation is also an important phenomenon in nonlinear acoustics. This is due to the nonlinear nature of the sound propagating medium such as the human tissues and the cracks in metals. They are especially useful in medical ultrasound imaging as it gives higher sensitivity and enables the detection of phenomena not seen in linear ultrasound imaging.

1.3   Phenomena in Nonlinear Acoustics

The following are common phenomena in nonlinear acoustics: sonic boom, acoustic levitation, musical acoustics and parametric arrays.

References

1. Tranquart, F., N. Grenier, V. Eder, and L. Pourcelot. 1999. Clinical use of ultrasound tissue harmonic imaging. *Ultrasound in Medicine and Biology* 25: 889–894.
2. Kristian, Haller. 2007. *Nonlinear Acoustics Applied to Nonodestructive Testing*, PhD. Thesis. Sweden: Blekinge Institute of Technology.

Chapter 2
Nonlinear Acoustic Wave Equations for Sound Propagation in Fluids and in Solids

2.1 Nonlinear Acoustic Wave Equations in Fluids

The usual nonlinear acoustics wave equations are meant for sound propagation in fluids (liquids and gases). They are originated from fluid mechanics. Hence the treatment will need the consideration of the following constitutive equations of fluid mechanics up to the second order of nonlinearity:

i.   Constitutive Equation of Nonlinear Acoustics in Fluids:

$$\frac{\partial \rho_a}{\partial t} + \rho_0 \nabla \cdot \vec{v_a} = -\rho_a \nabla \vec{v_a} \cdot \nabla \rho_a \tag{2.1}$$

ii. Equation of motion:

$$\rho_0 \frac{\partial \vec{v_a}}{\partial t} + \nabla p_a - \left(\xi + \frac{4}{3}\eta \right) \Delta \vec{v_a} = -\nabla L \tag{2.2}$$

iii. Equation of state:

$$\rho_a = \frac{p_a}{c_0^2} - \frac{B}{2A} p_a^2 / \rho_0 c_0^4 + \frac{\kappa}{\rho_0 c_0^4} \left(\frac{1}{c_v} - \frac{1}{c_p} \right) \frac{\partial p_a}{\partial t} \tag{2.3}$$

where κ = thermal conductivity, c_v, c_p = specific heat at constant volume and pressure, ξ = bulk viscosity, η = shear viscosity,

$\rho = \rho_0 + \rho_a$, $\vec{v} = \vec{v_0} + \vec{v_a}$, $p = p_0 + p_a$, L = Lagrangian energy density = $E_c - E_p = \frac{1}{2}\rho_0 v_a^2 - p_a^2 / 2\,\rho_0 c_0^2$.

Substitute (2.3) into (2.1) and get rid of ρ_0 and v_a, one obtains the following nonlinear propagation equation:

© Springer Nature Singapore Pte Ltd. 2021
W. S. Gan, *Nonlinear Acoustical Imaging*,
https://doi.org/10.1007/978-981-16-7015-2_2

$$\Delta p_a - \frac{1}{c_0^2}\frac{\partial^2}{\partial t^2}p_a + \frac{b}{c_0^2}\frac{\partial^3}{\partial t^3}p_a = -(\beta/\rho_0 c_0^4)\frac{\partial^2}{\partial t^2}p_a^2 - \left(\Delta + \frac{1}{c_0^2}\frac{\partial^2}{\partial t^2}\right)L \quad (2.4)$$

where $\beta = 1 + \frac{B}{2A}$ = nonlinear parameter, $b = (1/\rho_0 c_0^2)(\xi + \frac{4}{3}\eta) + (\kappa/\rho_0 c_0^2)(\frac{1}{c_v} - \frac{1}{c_p})$, ξ = bulk modulus, and η = shear viscosity.

2.1.1 The Westervelt Equation [1]

By deleting the last term on the right hand side of Eq. (2.4), one will obtain the Westervelt equation [1]:

$$\Delta p_a - \frac{1}{c_0^2}\frac{\partial^2}{\partial t^2}p_a + \frac{b}{c_0^2}\frac{\partial^3}{\partial t^3}p_a = -(\beta/\rho_0 c_0^4)\frac{\partial^2}{\partial t^2}p_a^2 \quad (2.5)$$

Westervelt equation can be applied in parametric acoustic array and in medical ultrasound.

2.1.2 The Burgers' Equation [2]

The Burgers' equation can be derived from the Navier Stokes equations which can be written as:

$$\nabla.\vec{v} = 0$$

$$\rho\frac{\partial v}{\partial t} + v\frac{\partial v}{\partial x} = 0 \quad (2.6)$$

and

$$(\rho v)_t + \nabla.(\rho\vec{v}\vec{v}) + \nabla p - \mu\nabla^2\vec{v} = 0 \quad (2.7)$$

where ρ = density, p = pressure, $\vec{v}$ = velocity, and μ = fluid viscosity.

Equations (2.6) and (2.7) are for divergence free incompressible flow given by (2.6) and $\rho_t = 0$ and the gravitational effects are neglected. By simplifying this to a one-D problem with propagation only in the x direction, one has.

$$\rho\frac{\partial v_x}{\partial t} + \rho v_x\frac{\partial v_x}{\partial t} + \rho v_y\frac{\partial v_x}{\partial y} + \rho v_z\frac{\partial v_x}{\partial z} + \frac{\partial p}{\partial x} - \mu\left(\frac{\partial^2}{\partial x^2}v_x + \frac{\partial^2}{\partial y^2}v_x + \frac{\partial^2}{\partial z^2}v_x\right) = 0.$$

There is no pressure gradient in one-D problem, then:

$$\rho \frac{\partial v_x}{\partial t} + \rho v_x \frac{\partial v_x}{\partial x} - \mu \frac{\partial^2}{\partial x^2} v_x = 0 \tag{2.8}$$

with ϵ = kinematic viscosity = $\frac{\mu}{\rho}$ and u = v_x, Eq. (2.8) becomes

$$u_t + uu_x = \epsilon\, u_{xx} \tag{2.9}$$

and with $\mu = 0$ for an inviscid fluid becomes

$$u_t + uu_x = 0 \tag{2.10}$$

which is the inviscid Burgers' equation.

Burgers' equation was proposed by Bateman [3] in 1915 and was subsequently analysed by Burgers' [2] in 1948. This partial differential equation is used in nonlinear acoustics, continuous stochastic processes, shock waves, heat conduction, turbulence, traffic flow, dispersive, water, and traffic flow. It is one of the very few nonlinear partial differential equations that can be solved exactly. It can be considered as a simplified form of Navier Stokes equations of fluid mechanics.

2.1.3 KZK Equation

KZK equation is commonly used in medical ultrasound. The KZK equation describes the propagation of finite amplitude sound waves in thermo-viscous medium. It was derived by Khokhlov and Zabolotskaya [3] and Kuznetsov [4]. KZK equation combines the effects of nonlinearity, diffraction, and absorption. There has been a substantial increase in the use of intense ultrasound in industrial and medical applications in recent years. Nonlinear effects due to intense ultrasound have become very important in many therapeutic and surgical procedures. It is of interest to know that the human body is a nonlinear medium. Ultrasound sources together with finite amplitude effects generate strong diffraction phenomena producing waveform varying from point to point within the sound beam. In addition, the significant absorption of sound due to the biological media also has to be considered.

Ultrasound sources together with finite amplitude effects generate strong diffraction phenomena producing waveform varying from point to point within the sound beam. The KZK equation for sound beam path in the z direction and the (x, y) plane perpendicular to that can be written as

$$\frac{\partial^2}{\partial z \partial \tau} p = \frac{c_0}{2} \nabla_{\perp}^2 p + \frac{\delta}{2c_0^3} \frac{\partial^3}{\partial \tau^3} p + \left(\beta/2\rho_0 c_0^3\right) \frac{\partial^2}{\partial \tau^2} p^2 \tag{2.11}$$

where p = acoustic pressure, $\tau = t - z/c_0$ = retarded time, c_0 = sound speed, ρ_0 = ambient density, β = nonlinearity coefficient $= 1 + \frac{B}{2A}$,

where A and B are the coefficients of the first and second order terms of the Taylor series expansion of the equation relating the material's pressure to its density and δ = sound diffusivity $= \frac{1}{\rho_0}(\frac{4}{3}\mu + \mu_B) + \frac{\kappa}{\rho_0}(\frac{1}{c_v} - \frac{1}{c_p})$ where μ = shear viscosity, μ_B = bulk viscosity, κ = the thermal conductivity, c_v and c_p = specific heat at constant volume and pressure respectively.

2.1.4 Nonlinear Acoustic Wave Equations for Sound Propagation in Solids

To describe sound propagation in solids, the elasticity equation has to be used. To describe the elastic behaviour of the material, the elastic energy is expressed as a function of strain by using the Taylor expansion of the elastic strain energy:

$$\rho_0 W = \frac{1}{2!} C_{ijkl} E_{ij} E_{kl} + \frac{1}{3!} C_{ijklmn} E_{ij} E_{kl} E_{mn} + \tag{2.12}$$

where C_{ijkl} = second order elastic constants, C_{ijklmn} = third order elastic constants, E = strain tensor, ρ = density, W = elastic strain energy.

For nonlinear acoustics in solids, the Lagrangian description is usually used. The acoustic equation of motion in solids in Lagrangian coordinates is given by

$$\rho_0 \frac{\partial^2}{\partial t^2} U = \nabla_a P_i \tag{2.13}$$

where P = Piola–Kirchhoff strain tensor and U = x − a = Lagrangian displacement or displacement relative to the natural position, a = natural or equilibrium position, and x = current coordinate of a particle and $F = \frac{\partial x}{\partial a}$ = deformation gradient tensor, E = Green's strain tensor with $E_{ij} = \frac{1}{2}(\frac{\partial U_i}{\partial a_j} + \frac{\partial U_j}{\partial a_i} + \frac{\partial U_k}{\partial a_i} \frac{\partial U_k}{\partial a_j})$ and

$$P_{ij} = C_{ijkl} \frac{\partial U_k}{\partial a_l} + \frac{1}{2} M_{ijklmn} \frac{\partial U_k}{\partial a_l} \frac{\partial U_m}{\partial a_n} + \frac{1}{3} M_{ijklmnpq} \frac{\partial U_k}{\partial a_l} \frac{\partial U_m}{\partial a_n} \frac{\partial U_p}{\partial a_q} \tag{2.14}$$

In (2.14), the second term accounts for quadratic nonlinearity and the third term the cubic nonlinearity and

$$M_{ijklmn} = C_{ijklmn} + C_{ijln}\delta_{km} + C_{jnkl}\delta_{im} + C_{jlmn}\delta_{ik} \tag{2.15}$$

where the first term accounts for material nonlinearity and the second, third, fourth terms the geometric nonlinearity or linear elastic constants.

From (2.13) one has

$$\rho_0 \frac{\partial^2}{\partial t^2} U_i = \frac{\partial}{\partial a_j} P_{ij} \tag{2.16}$$

In terms of (2.14), and 2.15), (2.16) becomes

$$\rho_0 \frac{\partial^2}{\partial t^2} U_i = \frac{\partial^2}{\partial a_j \partial a_l} U_k \left(C_{ijkl} + M_{ijklmn} \frac{\partial U_m}{\partial a_n} \right) \tag{2.17}$$

where M_{ijklmn} is given by (2.15).

Equation (2.17) is similar to the Westervelt equation for fluids.

References

1. Westervelt, P.J. 1963. Parametric acoustic array. *Journal of the Acoustical Society of America* 35: 535–537.
2. Burgers, J.M. 1948. A mathematical model illustrating the theory of turbulence. *Advance in Applied Mechanics* 1: 171–199.
3. Zabolotskaya, E.A., and R.V. Khokhlov. 1969. Quasi-plane waves in the nonlinear acoustics of confined beams. *Soviet Physics Acoustics* 8: 35–40.
4. Kuznetsov, V.P. 1971. Equations of nonlinear acoustics. *Soviet Physics Acoustics* 16: 467–470.

Chapter 3
Statistical Mechanics Approach to Nonlinear Acoustics

3.1 Introduction

In 1966, Gan [1] coined and invented the name transport theory in condensed matter physics during his Ph.D. works at the physics department of Imperial College London. Today transport theory is the foundation of the theoretical design of materials. Transport theory is also an important theory in statistical mechanics as it is closely related to phase transition. It is the theory of transport phenomena. In physics, transport phenomena are all irreversible processes of statistical nature stemming from the random continuous motion of molecules, mostly observed in fluids. Transport phenomena have wide applications. In condensed matter physics, the motion and interaction of electrons, holes and phonons are studied under transport phenomena. Transport phenomena are ubiquitous throughout the engineering discipline. Transport phenomena encompass all agents of physical change in the universe. Moreover, they are considered to be fundamental building blocks which developed the universe and which is responsible for the success of life on earth. The study of transport phenomena concerns the exchange of mass, energy charge, momentum and angular momentum between observed and studied systems. While it draws from fields as diverse as continuum mechanics and thermodynamics, it places a heavy emphasis on the commonalities between the topics covered. Mass, momentum, and heat transport all share a very similar mathematical framework. The following three topics are involved in transport phenomena:

1. Momentum transfer

The fluid is treated as a continuous distribution of matter. When a fluid is flowing in the x direction, parallel to a solid surface, the fluid has x directed momentum. By random diffusion of molecules, there is an exchange of molecules in the z direction. Hence the x-directed momentum has been transferred in z direction form the faster to the slower moving layer. The equation for momentum transport in Newton's law of viscosity:

© Springer Nature Singapore Pte Ltd. 2021
W. S. Gan, *Nonlinear Acoustical Imaging*,
https://doi.org/10.1007/978-981-16-7015-2_3

$$\tau_{zx} = -v\frac{\partial \rho v_x}{\partial z} \tag{3.1}$$

where τ_{zx} = flux of x directed momentum in the z direction, $v = \mu/\rho$ = momentum of diffusivity, z = distance of transport or diffusion, ρ = density, μ = dynamic viscosity.

2.   Mass transfer

When a system contains two or more components whose concentration varies from point to point, there is a natural tendency for mass to be transferred, minimizing any concentration difference within the system. Mass transfer in a system is governed by Fick's first law of diffusion which for a species A in a binary mixture consisting of A and B is given by

$$J_{AY} = -D_{AB}\frac{\partial C_a}{\partial y} \tag{3.2}$$

where D = diffusivity constant.

3.   Energy transfer

For the transfer of energy, the basic principle, the first law of thermodynamics which for a basic system is:

$$Q = -k\frac{dT}{dx} \tag{3.3}$$

where k = conductivity. This means the net flux of energy through a system equals the conductivity times the rate of change of temperature with respect to position.

3.2   Statistical Energy Analysis is Transport Theory

Statistical energy analysis (SEA) is based on the foundation of acoustic and vibration energies flow from high energetic to low energetic regions exactly as heat does in solids.

Statistical energy analysis (SEA) is the most famous method in acoustics and vibration and is a transport theory. Hence it is a part of statistical mechanics. This is a scaling up of the transport theory [1] in condensed matter physics from microscopic scales of molecules to macroscopic scales of acoustics and vibration. Here the concept of entropy and partition function from statistical mechanics are introduced into vibration. This is useful for the solving of a complex vibration system involving thousands of vibration modes and beyond the use of the method of fluid mechanics. SEA is a modal approach which provides physical insights into the mechanisms that govern energy flows in a complex acoustics and vibration system.

To consider SEA as a transport theory, one has to consider the thermodynamic parameters of entropy and partition function. Entropy is the product of the second law of thermodynamics. It is the measure of a system's thermal energy per unit temperature that is unavailable for doing useful work. Because work is obtained from ordered molecular motion, the amount of entropy is also a measure of the molecular disorder or randomness of a system.

Here one is concerned with system whose properties vary in space, raising the question of transport of particular wave energy between different parts of the structure. These system may take the form of discrete coupled subsystems or of continuous variation which is slow on some appropriate length scale. Examples from classical statistical mechanics are lined boxes of gas at different temperatures and the continuous generalisation of that problem to heat diffusion. In order to define a local temperature in the latter case, the spatial rate of change has to be sufficiently slow for local thermodynamics equilibrium to be assessed. Such problem occurs in many areas of physics and are treated by methods coming under general heading of transport theory usually based on the Boltzmann equation. Both discrete and continuous problems in this situation occur in the study of acoustics and vibration. There are a variety of problems concerning the spatial distribution of wave energy or intensity in a random medium, including the effects of the incoherent scattered wavefield. These are commonly treated by deriving suitable form of the Boltzmann transport equation, the approach being described variously on transport theory or radiative transfer theory. In all these cases, the aim is to obtain statistical information about the wavefield without actually solving the wave equation for individual realisation of the problem. The basic idea of statistical energy analysis (SEA) is to divide up a complex structure into a number of coupled subsystems and to model the energy flow between these in the spirit of transport theory, supposing it to mirror the way in which heat flows between coupled conductors. The concept of entropy is a precise way of expressing the second law of thermodynamics. It states that spontaneous change for an irreversible process in isolated system always proceeds in the direction of increasing entropy. SEA is a statistical method in vibroacoustics, entirely based on the application of energy balance that is the first law of thermodynamics.

The SEA method is by splitting a complex system into n elements. In all these subsystems, the vibration field is diffusive.,that is homogeneous and isotropic. The SEA approach consists in writing the exchange of energy between the sub-systems containing the vibrational energy E_i with $i = 1, \ldots n$. SEA describes the exchange of energy between these sub-systems. Within the sub-systems i, the vibrational energy is repartitioned among the N_i vibration modes. These modes carry vibrational energy E_i in the same ways atoms carry vibrational energy in solids and molecules carry kinetic energy in the kinetic theory of gases. In the two latter cases, the vibrational energy or kinetic energy is known as heat. In SEA the vibrational energy is equivalent to heat. The difference between the vibrational heat and true heat lies in the frequency of the underlying vibration. It is audio frequency range for the vibrational heat and thermal range for the true heat. Here the first law of thermodynamics which is concerned with energy balance will be introduced. The sub-systems receive energy from sources. Let the power being injected into the sub-system i be denoted P_{inj}^i. Vibrational energy will be dissipated by natural mechanisms such as attenuation of sound, damping of vibration, sounds absorption by walls etc. The power dissipated is denoted as P_{diss}^{i}. Besides this, the vibrational energy can be exchanged with adjacent sub-systems. Let the net exchanged power between sub-systems i and j be P_{ij}. In steady state condition, the energy balance will be given by

$$P_{diss}^{i} + \sum_{j \neq i} P_{ij} = P_{inj}^{i} \tag{3.4}$$

To show that statistical energy analysis is a transport theory and a part of statistical mechanics, the concepts of vibration entropy and vibration partition function are introduced below.

3.3   Statistical Energy Analysis

In statistical mechanics, the concept of macrostate and microstate will be introduced. The macrostate is characterized by E and N and the exact repartition on energy on modes will specify the microstate. Next step is to assess the number of microstates corresponding to a single macrostate. If one has Z energy quanta and N sites, then the number of possibilities to arrange Z among N is

$$W = \frac{(Z + N - 1)!}{Z!(N - 1)!} \tag{3.5}$$

For ordinary situations in acoustics, the number of energy levels Z is considerably larger than the number of modes N. Hence the sum $Z + N - 1$ differs from Z by a negligible term. Also $(Z + N - 1)! / Z!$ contains exactly $N - 1$ terms approximately equal to Z, (3.5) simplifies to

$$W = \frac{Z^{N-1}}{(N - 1)!} \tag{3.6}$$

One now introduces the Boltzmann's entropy of a system whose number of microstates is W, then

$$S = \text{entropy} = k_B \log W \tag{3.7}$$

where k_B = Boltzmann's constant.
Substituting (3.6) into (3.7) and with Stirling's approximation, $\log (N - 1)! = (N - 1) \log(n - 1) - (N - 1)$, one has

$$S = k_B (N - 1)[\log Z - \log(N - 1) + 1] \tag{3.8}$$

Replacing $(n - 1)$ by N and with $Z = E/\hbar\omega$, one has

$$S(E, N) = k_B N\left[1 + \log\left(\frac{E}{\hbar\omega N}\right)\right] \tag{3.9}$$

where $\hbar$ = Planck's constant.

This is the expression for the vibration entropy of a vibroacoustical system with energy E spread on N modes about the circular frequency ω.

The step one extends the vibration entropy to a nonlinear vibrating system. The nonlinear vibrating system produces a nonlinear energy cascade characterized by a power input to one frequency band, through random forcing or impact resulting in a response containing energy across a wider frequency range than the bandwidth of excitation. To extend the statistical energy analysis method to the energy cascade scenario, each of the subsystems has an assigned energy variable. Then nonlinear coupling loss factors (NCLFs) are derived to describe the energy flows between frequency bands. These NCLFs depend only on the properties of the nonlinear structure and not as a function of the forcing bandwidth or amplitude. The nonlinear equation of motion that governed a nonlinear multi-degree-of-freedom system consisting of a set of oscillators with nonlinear coupling up to the second order can be written as

$$\ddot{a}_j + 2\beta_j\omega_j\dot{a}_j + \omega_j^2 a_j = F_j + \sum_{k,l} C_{j,k,l} a_k a_l \tag{3.10}$$

where β_j, ω_j and F_j = natural frequency, damping factor and generalized forcing of the jth oscillator respectively, $C_{j,k,l}$ = tensor describing the strength of the second-order coupling between the oscillators.

Spelman and Langley [2] have derived an expression for the power balance as

$$P_r = \sum_s \sum_n G_{rsn} E_s E_n \tag{3.11}$$

Let the injected power for sources be P_i^{inj}. Then the vibrational energy δE supplied to the sub-system during time dt is $\delta E = P_i^{inj}$ dt. From here the increased rate of entropy by sources is

$$\frac{dS_i^{inj}}{dt} = \frac{P_i^{inj}}{T_i} \tag{3.12}$$

where P_i^{inj} = injected power of the sources and S = corresponding entropy.

For any thermodynamic system,

$$\frac{1}{T} = \left(\frac{\partial S}{\partial E}\right)_N \tag{3.13}$$

yielding

$$T = \frac{E}{kN} \tag{3.14}$$

Substitution of (3.14) in (3.12) then

$$\frac{dS^{inj}}{dt} = k\frac{P_i^{inj}}{E_i}N_i \tag{3.15}$$

Substitution of (3.11) in (3.12) and (3.15), produces

$$\frac{dS^{inj}}{dt} = \left[\sum_S \sum_n G_{rSn}E_S E_n\right]/T_i = k\left[\sum_S \sum_n G_{rSn}E_S E_n N_i\right]/E_i \tag{3.16}$$

3.4 Transport Theory Approach to Phase Transition

Phase transition is an important topic in statistical mechanics. There is singularity behaviour of transport properties such as specific heat, viscosity, Reynolds number during phase transition. The study of singularities of transport properties of the region surrounding the critical point of phase transition will enable understanding of phase transition. There are two interesting problems in nonlinear acoustics: turbulence and sonoluminescence which until today are still unsolved. Turbulence is a phase transition from laminar flow to turbulence flow. There is a tremendous huge increase in the Reynolds number or the viscosity tends to zero during the phase transition from laminar flow to turbulence flow. This is an example of the singularity behaviour of the transport properties: Reynolds number and viscosity. Besides this, there is spontaneous symmetry breaking (SSB) during the phase transition. The meaning of spontaneous symmetry breaking is explained as follows. In Gan [3] proposed turbulence as a second order phase transition with spontaneous symmetry breaking (SSB). Exampes of second order phase transition are magnetization,superconductivity and superfluidity in condensed matter physics. Hence turbulence is also a field in condensed matter physics. My hypothesis has been subsequently supported by the work of Goldenfeld [4]'s group which showed that turbulence has the same behaviour as magnetization. In his paper he presented experimental evidence that turbulence flows are closely analogous to critical phenomena from a reanalysis of friction factor measurements in rough pipe. He found experimentally two aspects that confirm that turbulence is similar to second order phase transition such as magnetization in a ferromagnet. These are experimentally verified power law scaling of correlation function which is reminiscent of the power law fluctuation on many length scales that accompany critical phenomena for example in a ferromagnet near its critical point which is second order phase transition. Another aspect is the phenomena of data collapse or Widom scaling [5]. For example, in a ferromagnet, the equation of state, nominally a function of two variables is expressible in terms of a single reduced variable which depends on a combination of external field and temperature. This has been confirmed by the experiments of Nikuradz [6] in 1932 and 1933 which showed data collapse. Goldenfeld [4]'s work proposed that the feature of the turbulence can be understood as arising from a singularity at infinite Reynolds number and zero roughness. Such singularities are known to arise as second order phase transition such as that occurs when iron is cooled down below the Curie temperature and becomes magnetic. This theory predicts that the small scale fluctuation in the fluid speed, a characteristic of turbulence are connected to the friction and can be demonstrated by plotting the data in a special way that causes all of the Nikuradze [8] curves at different roughness collapse into a single curve. According to Goldenfeld [4]'s study the formation of eddies in turbulence might be a similar phenomenon to the lining of spins in magnetization. Eddies are thus similar to the cluster of atoms. Goldenfeld et al. [6] hope that as a result of these discoveries, the approaches that solved the problem of phase transition will now find a new and unexpected application in providing a fundamental understanding of turbulence. In Gan [7] extended Landau [8]'s theory of the second order phase transition from phenomenology to a more rigorous approach by using statistical mechanics and gauge theory. We propose turbulence as a classical analog of Bose-Einstein condensation and the Gross-Pitaevskii equation is used to derive the condensate free energy. The critical value of the order parameter, the condensation wave function is determined.This is the value when turbulence occurs and SSB in the ground state of the condensation free energy takes place. Being a condensate, there is molecular pairing in turbulence. We determine the numerical value of the condensate free energy with the use of the coupled oscillation model for the pair of molecules.

Our expression for the condensation free energy also yields a power series in terms of the order parameter in agreement with the Landau phenomenology of second order phase transition. This confirms that turbulence is a condensate since Gross-Pitaevskii equation is the equation for condensate. We conclude that our understanding of turbulence is a second order phase transition and is a condensate with molecular pairing.

References

1. Gan, Woon Siong. 1969. *Transport Theory in Magnetoacoustics*, PhD Thesis. Imperial College London.
2. Spelman, G.M., and R.S. Langley. 2015. Statistical energy analysis of nonlinear vibrating system. *Philosophical Transactions of the Royal Society A: Mathematical, Physical and Engineering Sciences* A373: 20140403.

3. Gan, Woon Siong. 2009. Turbulence is a second order phase transition with spontaneous symmetry breaking. In *Proceedings of the ICSV16*. Krakow, Poland.
4. Goldenfeld, N. 2006. Roughness-induced critical phenomena in a turbulent flow. *Physical Review Letter* 96: 044503.
5. Widom, B. 1965. Equation of state in the neighbourhood of critical point. *The Journal of Chemical Physics* 43: 3898–3905.
6. Nikuradze, J. 1933. *VDI, Forschungsheft, Stromungsgesetze in rauhen Rohren (Laws of flow in rough pipes), NACA Technical Memorandum 1292*, vol. 361. Berlin: Springer-Verlag.
7. Gan, Woon Siong. 2013. Towards the understanding of turbulence. *Journal of Basic and Applied Physics* 12 (5): 1–9.
8. Landau, L. 1937. Theory of phase transformation, I, Theory of phase transformation, I. Zh.Eksp.Teor.Fiz. **7**: 19, 627.

Chapter 4
Curvilinear Spacetime Applied to Nonlinear Acoustics

4.1 Introduction and Meaning of Curvilinear Spacetime

It is an intrinsic property of nonlinear acoustics that the acoustic field path is nonlinear or curvilinear unlike linear acoustics with linear wavefield paths. This is due to the quadratic and higher order nonlinearity nature of the nonlinear acoustic wave equation which is dealing with high amplitude or high intensity sound fields. Hence the curvilinear coordinates system has to be used to describe the curvilinear field paths. An example is the use of curvilinear spacetime by Einstein [1] to describe the curvilinear paths of the gravitational force.

In curvilinear spacetime, the curvilinear coordinates system has to be used. These are a coordinate system for Euclidean space in which the coordinate lines may be curved. Curvilinear coordinates may be derived from a set of Cartesian coordinates by using transformation that is locally invertible at each point. That is one can convert a point given in a Cartesian coordinates system to its curvilinear coordinates and back. The formalism of curvilinear coordinates provides a unified and general description of the standard coordinate system. Well-known examples of curvilinear coordinates system in three-dimensional Euclidean space are cylindrical and spherical coordinates.

Curvilinear spacetime means the extension of the curvilinear coordinates system from three dimension to four dimension space with the inclusion of time as the fourth coordinate in the four-dimensional spacetime continuum.

Cartesian coordinates is a special case of curvilinear coordinates when the path is a straight line. It arises from the definition of the line element and arc length. The definition of the line element and arc length arises from geometry. The line element is a line segment associated with an infinitesimal displacement vector in a metric space. The length of this line element which is a differential arc length is a function of the metric tensor and is denoted by ds. The coordinate-independent definition of the square of the line element ds is the square of the length of an infinitesimal displacement dq whose square root is used for computing curve length:

© Springer Nature Singapore Pte Ltd. 2021
W. S. Gan, *Nonlinear Acoustical Imaging*,
https://doi.org/10.1007/978-981-16-7015-2_4

$$ds^2 = dq.dq = g(dq, dq) \tag{4.1}$$

where g = metric tensor, denotes inner product and dq = infinitesimal displacement on the Riemannian manifold. The n-dimensional general curvilinear coordinates q $= (q^1, q^2, q^3, \ldots q^n)$ and

$$ds^2 = g_{ij}dq^i dq^j = g \tag{4.2}$$

The following is an example of the line element in Euclidean space. The simplest example is the Cartesian coordinates. Hence the metric tensor is just the Kronecker delta:

$$g_{ij} = \delta_{ij} \tag{4.3}$$

Here i, j = 1, 2, 3 for space and in matrix form gives

$$[g_{ij}] = \begin{pmatrix} 1 & 0 & 0 \\ 0 & 1 & 0 \\ 0 & 0 & 1 \end{pmatrix} \tag{4.4}$$

And the general curvilinear coordinates reduces to Cartesian coordinates:

$$(q^1, q^2, q^3) = (x, y, z) \Rightarrow dr = (dx, dy, dz). \tag{4.5}$$

and

$$ds^2 = g_{ij}dq^i dq^j = dx^2 + dy^2 + dz^2.$$

4.2   Principle of General Covariance

All fundamental equations of physics have covariance principle. That is, they all remain the same for under different coordinates systems. This is a gauge invariance property of the equations. In nonlinear acoustics, high intensity sound fields are used which have curvilinear paths and curvilinear coordinates have to be used. This is a description of the propagation of nonlinear sound fields in curvilinear spacetime. The usual equation used for nonlinear acoustics, the KZK equation is actually an approximate form of the Navier- Stokes equations. It is meant for paraxial beams of sound waves in thermos-viscous fluids. In our work, our treatment of high intensity soundfields in curvilinear spacetime will be without this approximation.

One starts with an introduction to the four-dimensional space-time continuum:

$$ds^2 = -dX_1^2 - dX_2^2 - dX_3^2 + dX_4^2. \tag{4.6}$$

where ds is the magnitude of the linear element pertaining to points of the four-dimensional continuum in infinite proximity. If ds is positive, it is time-like and if it is negative, it is space-like. ds^2 has a value which is independent of the orientation of the local system ofcoordinates. Representing dX_v by definite linear homogeneous expression then

$$dX_v = \sum_\sigma a_{v\sigma} dx_\sigma \tag{4.7}$$

Substituting these expressions in (4.1), then

$$ds^2 = \sum_{\tau\sigma} g_{\sigma\tau} dx_\sigma dx_\tau \tag{4.8}$$

The $g_{\sigma\tau}$ is known as the fundamental tensor.

4.3   Contravariant and Covariant Four-Vectors

For any space–time coordinates $x_1, x_2, x_3, x_4, g_{\sigma\tau}$ will be function of space-and time. The coordinates will be curvilinear coordinates describing curvilinear non-uniform motion. The spacetime territory in question is under the sway of a gravitational field which generates the accelerated motion of the bodies relatively to K.

To implement the concept of general covariance, tensors have to be used to represent the coordinates transformation. The components of tensors will be functions of the system of coordinates. These components can be calculated for the new systems of coordinates if they are known for the original systems of coordinates using certain rules of transformation. The equations of transformation for the tensor components are linear and homogeneous. The laws of the formation of tensors are linear and homogeneous. The laws of the formation of tensors will require covariant laws.

The tensor components are known as contravariant four-vector components dx_v with the transformation law given by

$$dx_\sigma = \sum_v \frac{\partial}{\partial x_v} x\sigma \, dx_v \tag{4.9}$$

The dx_σ' are linear and homogeneous functions of dx_v. In general, contravariant four-vector components can be represented by:

$$A'^\sigma = \sum_v \frac{\partial}{\partial x_v} x_\sigma A^v \tag{4.10}$$

A contravariant four-vector also means anything which is defined relatively to the system of coordinates by four quantities A^v and which is transformed by the same law given by (4.10). From (4.10) it follows that A^σ are also components of a four-vector.

The four quantities will become the components of a covariant four-vector if for any arbitrary choice of the contravariant four-vector B^v,

$$\sum_v A_v B^v = \text{Invariant} \tag{4.11}$$

From (4.11), the law of transformation of a covariant four-vector will follow:

$$\sum_\sigma A'_\sigma B'^\sigma = \text{Invariant} = \sum_v A_v B^v \tag{4.12}$$

If one replaces the one on the right hand side of (4.12) by the equation resulting from the inversion of (4.10), then

$$\sum_\sigma B'^\sigma \sum_v \frac{\partial x_v}{\partial x'_\sigma} A_v = \sum_\sigma B'^\sigma A'_\sigma \tag{4.13}$$

Since (4.13) is true for arbitrary values of B'^σ, the law of transformation of a covariant fourvector will be given by

$$A'_\sigma = \sum_v \frac{\partial}{\partial x'_v} x_v A_v \tag{4.14}$$

Next is to derive the geodesic equation to understand the equation of motion of a single particle or Newton's equation of motion in curvilinear space–time. Then add on stress-stain relation on curvilinear space–time.

4.4 Contravariant Tensors and Covariant Tensors

The product $A^{\mu v}$ of the two contravariant four-vectors A^μ and B^v is given by

$$A^{\mu v} = A^\mu B^v \tag{4.15}$$

which consists of sixteen products. Then from (4.10) and (4.14), a law of transformation will be:

$$A'^{\sigma \tau} = \frac{\partial x'_\sigma}{\partial x_\mu} \frac{\partial x'_\tau}{\partial x_v} A^{\mu v} \tag{4.16}$$

which is known as a contravariant tensor of the second rank...The contravariant four-vector will be known as the contravariant tensor of the first rank.

Similarly taking $A_{\mu\nu}$ as a product of two convariant four-vectors A_μ and B_ν,

$$A^{\mu\nu} = A^\mu B^\nu \qquad (4.17)$$

which consists of sixteen products. Then from (4.10) and (4.14), a law of transformation will be:

$$A'^{\sigma\tau} = \frac{\partial x'_\sigma}{\partial x_\mu} \frac{\partial x'_\tau}{\partial x_\nu} A^{\mu\nu} \qquad (4.18)$$

which is known as a contravariant tensor of the second rank...The contravariant four-vector will be known as the contravariant tensor of the first rank.

Similarly taking $A_{\mu\nu}$ as a product of two convariant four-vectors A_μ and B_ν,

$$= A_\mu B_\nu \qquad (4.19)$$

It follows that the law of transformation will be given by

$$A'^{\sigma\tau} = \frac{\partial x_\mu}{\partial x'_\sigma} \frac{\partial x_\nu}{\partial x'_\tau} A^{\mu\nu} \qquad (4.20)$$

Equation (4.20) defines the covariant tensor of the second rank and a scalar is known as a covariant tensor of zero rank. Examples of the outer multiplication of tensors of the first rank (vectors) are given by Eqs. (4.17) and (4.19).

4.5   The Covariant Fundamental Tensor $g_{\mu\nu}$

We denote $g_{\mu\nu}$ as a fundamental covariant tensor. We denote $g_{\mu\nu}$ as fundamental covariant tensor. It has the symmetrical property of $= g_{\nu\mu}$. The square of the linear element ds can be expressed as

$$ds^2 = g_{\mu\nu}dx_\mu dx_\nu \qquad (4.21)$$

The covariant tensor $g^{\mu\nu}$ is formed by taking the co-factor of each of the $g_{\mu\nu}$ elements and divide it by the determinant $g = |g_{\mu\nu}|$. It is denoted by $g^{\mu\nu} = g^{\nu\mu}$. Using a known property of determinant, we have

$$g_{\mu\sigma} g^{\nu\sigma} = \delta^\nu_\mu \qquad (4.22)$$

where δ_μ^v denotes 1 or 0, according to $\mu = v$ or $\mu \neq v$. Hence ds^2 can be written as

$$ds^2 = g_{\mu\sigma}\delta_v^\sigma \, dx_\mu dx_v \tag{4.23}$$

Using the following rule for the multiplication of determinants:

$$\left|g_{\mu a}g^{av}\right| = \left|g_{\mu a}\right| \times \left|g^{av}\right|$$
$$\left|g_{\mu a}g^{av}\right| = \left|\delta_\mu^v\right| = 1$$

One obtains

$$\left|g_{\mu a}\right| \times \left|g^{av}\right| = 1 \tag{4.24}$$

4.6 Equation of Motion of a Material Point in the Gravitational Field

According to the general theory of relativity, the equation of motion of the material point in the gravitational field is given by

$$\frac{d^2}{ds^2} = \Gamma_{\mu v}^\tau \frac{dx_\mu}{ds} \frac{dx_v}{ds} \tag{4.25}$$

with $\Gamma_{\mu v}^\tau = -\{\mu v, T\}$.

This is a covariant system of equations. $\Gamma_{\mu v}^\tau$ an component of the gravitational field and they provide the condition of the deviation of the motion from uniformity. When they vanish, the point moves uniformly in a straight line.

4.7 The Laws of Momentum and Energy for Matter, as a Consequence of the Gravitational Field Equations

The general field equations of gravitation in mixed form are given by

$$\frac{\partial}{\partial x_\alpha} + \Gamma_{va}^\beta = -\kappa\left(T_{\mu v} - \frac{1}{2}g_{\mu v}T\right) \tag{4.26}$$

$$\sqrt{-g} = -1$$

where t_σ^α = energy component of the gravitational field.

If (4.26) is multiplied by $\partial g^{\mu\nu}/\partial$ and vanishing of $g_{\mu\nu}\frac{\partial g^{\mu\nu}}{\partial x_\sigma}$, one has

$$\frac{\partial A_\sigma^\alpha}{\partial x_\alpha} + \frac{1}{2}\frac{g^{\mu\nu}}{\partial x_\sigma}T_{\mu\nu} = 0 \tag{4.27}$$

and with $\partial(t_\mu^\sigma + T_\mu^\sigma)/\partial x_\sigma = 0$ gives

$$\frac{\partial T_\sigma^\alpha}{\partial x_\alpha} + \frac{1}{2}\frac{g^{\mu\nu}}{\partial x_\sigma}T_{\mu\nu} = 0 \tag{4.28}$$

where t_σ^α = energy component of the gravitational field.
Equation (4.28) can be rewritten as

$$\frac{\partial T_\sigma^\alpha}{\partial x_\alpha} = -\Gamma_{\alpha\sigma}^\beta T^{\alpha\beta} \tag{4.29}$$

The right hand side of (4.29) shows the energetic effect of the gravitational field on matter.

4.8 The Euler Equation of Fluids in the Presence of the Gravitational Field

Here we will derive the Euler equation of fluids in the presence of the gravitational field. Let P = pressure, ρ = fluid density. We write the contravariant symmetrical tensor as the contravariant energy tensor of the fluid given by

$$T^{\sigma\beta} = -g^{\alpha\beta}P + \rho\frac{dx_\sigma}{ds}\frac{dx_\beta}{ds} \tag{4.30}$$

The corresponding covariant tensor will be given by

$$T_{\mu\nu} = g_{\mu\nu}P + g_{\mu\alpha}g_{\mu\beta}\frac{dx_\alpha}{ds}\frac{dx_\nu}{ds}\rho \tag{4.31}$$

The corresponding mixed tensor is given by

$$T_\sigma^\alpha = -\delta_\sigma^\alpha P + \frac{dx_\beta}{ds}\frac{dx_\alpha}{ds}\rho \tag{4.32}$$

Inserting the right hand side of (4.32) into (4.29), one will obtain the Euler equation of fluids in in the presence of the gravitational field. Equation (4.29) consists of four equations together with the equation of relation between P and ρ plus the following equation:

$$g_{\alpha\beta}\frac{dx_\alpha}{ds}\frac{dx_v}{ds} = 1 \tag{4.33}$$

with $g_{\alpha\beta}$ being given are sufficient to solve the six unknowns P, ρ, $\frac{dx_1}{ds}$, $\frac{dx_2}{ds}$, $\frac{dx_3}{ds}$, $\frac{dx_4}{ds}$. If $g_{\mu\nu}$ is un-known, then with (4.26), there will be eleven equations for defining the ten functions $g_{\mu\nu}$ given as

$$\frac{\partial}{\partial x_\alpha}\left(-\mathrm{P} + g_{\alpha\beta}\frac{dx_\beta}{ds}\frac{dx_\alpha}{ds}\rho\right) = -\Gamma^\beta_{\alpha\sigma}T^\alpha_\beta \tag{4.34}$$

4.9   Acoustic Equation of Motion for an Elastic Solid in the Presence of Gravitational Force

The action principle using the method Lagrange multipliers will be used to derive the acoustic equation of motion for elastic solids in a curvilinear space–time in the presence of the gravitational field. First the Einstein equations for the gravitational field potentials, viz. the components of the metric tensors on the space–time manifold can be derived from an action principle with a scalar density Lagrangian function given by $(-g)1/2R$, where g = fundamental tensor and R = scalar curvature of the space–time manifold. The action principle leads to the field equations

$$G_{ik} = -\frac{1}{2}\mathrm{R}\,g_{ik} = 0 \tag{4.35}$$

where G_{ik} = Einstein tensor and R_{ik} = Ricci tensor. Equation (4.35) are the field equations valid in matter and field free space-time.

In order to obtain the field equations in the presence of matter, the action principle must be modified by the addition to the Lagrangian of a term describing the matter present. The fundamental in the assumption made is that the action principle in the general case shall have the form:

$$\delta A = 0 \tag{4.36}$$

where A is the action given by

$$A = \int\left\{-(-g)^{1/2}\frac{R}{2\kappa} + \mathrm{L}\right\}d^4v \tag{4.37}$$

In (4.37), κ = gravitational constant, L = a scalar density function describing the matter and fields present, and the integral is taken over a fixed region of the space–time manifold. The variations of the dynamical variables considered in (4.36) are such that they vanish on the boundary of the region of integration. They also be subjected to constraint implied by the existence of geometric laws which have not been satisfied identically by means of a potential representation. From (4.36) and (4.37), follow the modified gravitational field equation

$$(-g)^{1/2} G_{ik} + \kappa T_{ik} = 0 \qquad (4.38)$$

where T_{ik} = symmetric tensor density given by

$$T_{ik} = 2 \frac{\delta L}{\delta g_{ik}} \qquad (4.39)$$

$\frac{\delta L}{\delta g_{ik}}$ is the Lagrangian derivative of L with respect to the gik. T^{ik} is called the stress-energy-momentum tensor of the matter and fields represented by the term L in the Lagrangian denity. Besides the gravitational field Eq. (4.38), (4.36) also implies field equation for the other variables occurring in L. The identity required by the principle of general covariance are in this case the identity following the Bianchi identity

$$\nabla_k G^{ik} = 0$$

viz they are

$$\nabla_k T^{ik} = 0 \qquad (4.40)$$

∇_k represents covariant differentiation with respect to x^k based on the Christoffel symbols defined in terms of the metric tensors.

Next is to use the action principle to derive the acoustic equation of motion for an elastic solid valid in general relativity, but using the equivalent to the Lagrangian description in nonrelativistic continuum mechanics. The Eulerian descriotion is used here.

The motion of a material continuum can be described by giving the three scalar fields X^K. However, to be able to use analogies with non-relativistic theories, it is preferable to express L in terms of the world velocity w_i and the mass density ρ, in addition to any other variables required. It is then followed that the geometric laws connecting these quantities must be taken into account. These laws, expressing the unit magnitude of w^i, the constancy of along a world line and the conservation of mass.

$$g_{ik} w^i w^k + 1 = 0 \qquad (4.41)$$

$$w^i \partial_i X^K = 0 \qquad (4.42)$$

$$(\rho w^i) = 0 \tag{4.43}$$

The scalar density L, assumed to be a function of the g_{ik}, w^i, ρ, X^K, and $\partial_i X^K$ is constructed in such a way that the tensor density $T^{ik} = 2\, \delta L/\delta g_{ik}$ should reduce to the Lorentz invariant form of the stress-energy–momentum tensor of an elastic solid in a flat space–time.

The stress = energy–momentum tensor of an elastic solid can now be calculated according to (4.39),

$$T^{ik} = \partial \frac{\delta L}{\delta g_{ik}} = \rho(1 + \Sigma)w^i w^k - t^{ik} \tag{4.44}$$

where $\Sigma = \Sigma(c^{-1\,KL})$, and the elastic strain density t^{ik} is given by

$$t^{ik} = -2\rho \frac{\partial \Sigma}{\partial C^{-1KL}} \partial_m X^K \partial_n X^L g^{mi} g^{nk} \tag{4.45}$$

and $g_{ik}\, w^i w^k$ has been put equal to 1 after differentiation. c^{-1} is a scalar given by

$$c^{-1KL} = g^{k1} \partial_k X^K \partial_1 X^L \tag{4.46}$$

In terms of the covariant derivative, one has

$$\nabla_k T_i^k = 0 \tag{4.47}$$

Equation (4.47) together with (4.42) and (4.43) constitute a set of $4 + 3 + 1 = 8$ equations for eight unknowns w^i, ρ, and X^K, i.e. they are a complete system for the acoustic equation of motion for an elastic solid in curvilinear space–time. At first sight this appears to contradict the principle of general covariance, but note that g_{ik} have been implicitly taken as known in the above discussion, whereas actually they are to be determined from the Einstein Eq. (4.38) which satisfy the Bianchi identities. Equation (4.47) are thus in a sense redundant, since they are implied by the Bianchi identities. On the other hand, the fact that (4.47) follows from the action principle independently of the Einstein equations provides a check on the consistency of the whole theory.

Reference

1. Kox, A.J., M.J. Klein, and R. Schulmann, eds. 1997. *The Collected Papers of Albert Einstein*, vol. 6. Princeton University Press.

Chapter 5
Gauge Invariance Approach to Nonlinear Acoustical Imaging

5.1 Introduction

Electron–phonon interaction is an important topic in nonlinear acoustics. It is a transport theory. It is necessary for the basic understanding of sound propagation in solids. The usual treatment is by the use of manybody perturbation theory. Perturbation theory is an important method for describing real quantum systems because it is usually very difficult to find exact solutions to the Schrödinger equation for Hamiltonian of even moderate complexity. Usually the Hamiltonians with exact solutions are only for idealized systems which are inadequate to describe most systems.

Hence with perturbation theory, one can use the known solutions of these simple Hamiltonians to generate solutions for more complex and realistic systems. This is done by adding a small term to the mathematical description of the exactly solvable problem.

For the use of the manybody perturbation theory to describe electron–phonon interaction, usually there is the expansion of the Green's function in powers of the interaction. The many particle Green's function is introduced. The one particle Green's function is non-interacting. One can derive an exact explicit expression for the n-particle Green's function in terms of the one-particle Green's function. This forms the basis of manybody perturbation theory.

There are limitations to the manybody perturbation theory. It is only limited up to the second order perturbation. Beyond that infinity occurs.

5.2 Gauge Invariance Formulation of Electron–Phonon Interaction

Woon Siong Gan [1] introduced gauge invariance approach to acoustic fields. The gauge invariance formulation was used for the electromagnetic waves. Maxwell's equations is the first display of gauge invariance. Gauge invariance is the basis of the

© Springer Nature Singapore Pte Ltd. 2021
W. S. Gan, *Nonlinear Acoustical Imaging*,
https://doi.org/10.1007/978-981-16-7015-2_5

Yang Mills theory [2] which is now the foundation of the Standard Model of particle physics. It has the advantage that the infinity problem of the manybody perturbation theory is removed.

Electron–phonon interaction is ubiquitous in condensed matter physics. It can also be considered as the low energy regime of the high energy physics. It will be treated as a consequence of local gauge invariance. It will be a study of the motion of electrons in a solid with elastic properties. This will be mutual interactions resulting from the local symmetry properties. The elementary excitation of the solid will be described by the gauge fields.

The elastic fields will be introduced as gauge fields. This is a consequene of the invariance under local space-time translation of the condensed matter system. The outcomes are: (i) the coupling of electrons with the gauge field generated straight-forwardly with the electrons scattered by the elastic fluctuations. This interaction is identified with the electron-phonon coupling. (ii) the spin-orbit effects produced by the electron dynamics as a consequence that the elastic field couples with the spin current will induce novel magneto-elastic phenomena as a reflection of gauge symmetries. This can find relevance in the field of spintronics.

To develop the gauge invariance formulation of the electron-phonon interaction, first the gauge invariance is imposed under general space-time dependent trans-lations. The electronics system is described by a simple Lagrangian density with a free-electron structure. Local symmetry necessitates the introduction of a gauge field associated with the elastic properties of the solid. In this gauge invariance treatment, the phonon field is a gauge field. The invariance of the electron Lagrangian density is imposed under coordinate dependent translations. Infinitesimal transformation of this kind involves rigid-body displacements of constant translations and deforma-tions with an associated strain field. We consider non-relativistic theory as the first approximation. In this case, the Lagrangian density describing a non-relativistic elec-tron moving in the crystalline lattice, in units of $\hbar = 1$, where $\hbar$ = Planck's constant is given as

$$L = i\psi^+ \frac{\partial \psi}{\partial t} - \frac{1}{2m^*} \nabla \psi^+ \cdot \nabla \Psi \qquad (5.1)$$

where ψ = Pauli spinor describing the electron, and m* = effective mass,. The Lagrangian density of (5.1) describes a free electron in the lattice.

In the collective excitation of a crystalline lattice, there is a symmetry breaking from a continuous translational symmetry to a discrete subgroup, corresponding to translation between equivalent points in different primitive cells. The invariance of the Lagrangian density in Eq. (5.1) has to be subjected to a local gauge transformation. The discrete nature of the crystalline translation can be shown by the Bloch theorem which ensures that a translation x $\rightarrow$ x + a where a is an appropriate lattice translation vector acts on the electron wave function ψ by means of the translation operator $\tau(a)$ = exp[− ia.p] where p = i∇= momentum operator.

$$\Psi' = (a)\psi = \exp[-ia.p]\psi \tag{5.2}$$

where ψ' = transformed wave function, different from ψ only by a phase factor $\exp[-i\vec{k}.\vec{a}]$. Here a gauge field is associated to slight distortion of the crystal lattice, inducing a space–time dependent translation of the electron wavefunction.

With the translation operator depending on the space-time coordinates, the transformation status will be local gauge symmetry. The local gauge transformation can be represented by

$$\Psi' = \tau[\delta a(x, t)]\psi = [1 - i\delta a(x, t).p]\psi \tag{5.3}$$

where ψ = Pauli spinor, $\delta a(x, t)$ = infinitesimal translation, a = lattice vector, and τ = translation operator. This is instead of a translation by a lattice vector a, the Pauli spinor ψ is subjected to an infinitesimal translation $\delta a(x, t)$ in space-time.

For the implementation of local gauge invariance, the ordinary derivative $\partial_\mu = \partial/\partial x^\mu$ has to be replaced by the covariant derivative

$$D_\mu = \partial_\mu + gW_\mu \tag{5.4}$$

where g = coupling constant and W_μ = vector potential is a key gauge field. It can be written as infinitesimal generator of space–time translation as,

$$W_\mu = -i\,R_{\mu\nu}p^\nu \tag{5.5}$$

where $p^\nu = i\,\partial^\nu$ = four-momentum operator, and $R_{\mu\nu} = R_{\nu\mu}$ is asymmetric second rank tensor.

The electron–phonon interaction is represented by local gauge invariance. For sound propagation in solids, $R_{\mu\nu}$ are equivalent to the tensor components of the elasticity field.

In order to define the covariant derivative D_μ, one needs to know the value of the gauge field W_μ at all space–time points. Instead of manually specifying the values of this gauge field, it can be given as the solution to a field equation. In order for the Lagrangian that generates this field equation to be locally invariant, one needs the following form for the gauge field Lagrangian:

$$L_{GF} = -\frac{1}{4}F_{\mu\nu\beta}F^{\mu\nu\beta} = -\frac{1}{4}\text{Tr}[F_{\mu\nu} \neq^{\mu\nu}] \tag{5.6}$$

where $F_{\mu\nu}$ being the elasticity fields.

The elasticity fields can be obtained from the commutator of the covariant derivative as [3]:

$$GF_{\mu v} = i[D_{\mu}, D_{\delta}] = i(D_{\mu}D_v - D_v D_{\mu})$$
$$= ig(\partial_{\mu} W_v - \partial_v W_{\mu}) + ig^2[W_{\mu}, W_v]$$

$$= g\, F_{\mu v \beta}\, p^{\beta} \tag{5.7}$$

The symbol Tr in Eq. (5.6) means taking the trace over the vector space of the field. The commutator $[W_{\mu}, W_{\nabla}]$ does not vanish. The gauge field of electron–phonon interaction is non-abelian. It is also invariant under the gauge transformation.

$$R'_{\mu v} = R_{\mu v} - \partial_{\mu} a_v - \partial_{\delta} a_{\mu} \tag{5.8}$$

where a_{μ} is an arbitrary four-vector.

Then the Lagrangian density of the elasticity fields will become

$$L_{GF} = -\frac{1}{2}\left[\partial_{\mu} R_{v\beta}\partial^{\mu} R^{v\beta} - \partial_{\mu} R_{v\beta}\partial^v R^{\mu\beta}\right] \tag{5.9}$$

As a simplification, one considers only translations in space coordinates. This corresponds to the following transformation:

$$x' = x + \delta\alpha(x, t), \quad t' = t \tag{5.10}$$

Then the covariant derivatives will simplify to:

$$D_t = \frac{D}{Dt} = \frac{\partial}{\partial t} - igc_s \vec{R}.\vec{p} = \frac{\partial}{\partial t} - ig_0 \vec{R}.\vec{p} \tag{5.11}$$

and

$$D_i = \frac{D}{Dx^i} = \frac{\partial}{\partial x_i} - ig\vec{R_i}.\vec{p} \tag{5.12}$$

where $\vec{R}$ and $\vec{R_i}$ are gauge vector potentials related to the phonon and strain fields, $g_0 = c_s\, g$ and g are the corresponding coupling constants, and $c_s =$ sound velocity in the solids.

The full gauge invariant Lagrangian density is given by

$$L = i\psi^+ \frac{D\psi}{Dt} - \frac{1}{2m^*}(D_i^+\psi^+)(D_i\psi) + L_{GF} \tag{5.13}$$

All components of the phonon vector field are coupled phonons will be scattered by fluctuation of the strain field.

5.3 Illustration by a Unidirectional Example

For further simplification, one takes the case of a one-dimensional space. Then the Lagrangian density will become

$$L = i\psi^+ \frac{\partial \psi}{\partial t} - \psi^+ g_0 R \frac{\partial \psi}{\partial x}$$
$$- \frac{1}{2m^*}[(\partial_x - g\, R_x \partial_x)\psi^+][(\partial_x - gR_x\partial_x)\psi] \qquad (5.14)$$

Here R is the off-diagonal component of the tensor $R_{\mu\nu}$, with $R = R_{01} = R_{10}$ and $R_x = R_{11}$. Then the Lagrangian density of the elasticity becomes

$$L_{GF} = -\frac{1}{2}[(\partial_0 R_x)^2 + (\partial_x R)^2 - (\partial_0 R)^2 - 2(\partial_x R)(\partial_0 R_x)] \qquad (5.15)$$

yielding the Hamiltonian density of the gauge field as

$$H_{GF} = \frac{1}{2}\left[\frac{1}{c_s^2}\left(\frac{\partial R}{\partial t}\right)^2 + \left(\frac{\partial R}{\partial x}\right)^2 - \frac{1}{c_s^2}\left(\frac{\partial R_x}{\partial t}\right)^2 \right] \qquad (5.16)$$

R is identified with the phonon field ϕ, with longitudinal polarization for the one-dimensional line. The Hamiltonian density can be rewritten as

$$H_{FG} = \frac{1}{2}\left[\frac{1}{c_s^2}\left(\frac{\partial \phi}{\partial t}\right)^2 + \left(\frac{\partial \phi}{\partial x}\right)^2 \right] \qquad (5.17)$$

producing the linear acoustic wave equation with wave velocity c_s...

5.4 Quantization of the Gauge Theory

The above treatment is classical gauge theory for electron–phonon interaction. To quantize the gauge theory, the phonon variables will be introduced. The phonon field, the displacement operator will be expanded in terms of the creation and annihilation operators as a_q^+ and a_q as follows [4]:

$$\phi(x, t) = \sum_q \sqrt{\frac{1}{2L\rho W_q}}\left[a_q e^{i(qx - W_q t)} + a_q^+ e^{-i(qx - W_q t)} \right] \qquad (5.18)$$

where ρ = mass density, W_q = energy dispersion relation, L = size of the system, and $\phi(x, t)$ = displacement operator.

The second quantized Hamiltonian can be written as

$$\widehat{H}_{FG} = \int ds\, H_{FG} = \sum_q W_q a_q^+ a_q \tag{5.19}$$

When the electron field is second-quantized, the standard electron–phonon interaction can be obtained. Expanding the electron field in plane waves, one has.

$$\Psi = \sum_{k,\sigma} \frac{1}{\sqrt{L}}\left(e^{ikx} c_{k\sigma} + e^{-ikx} c_{k\sigma}^+\right) \tag{5.20}$$

where $(c_{k\sigma}, c_{k\sigma}^+)$ are fermion destruction and creation operators of particles with wave number k and spin σ.

5.5   Coupling of Elastic Deformation with Spin Currents

The gauge invariance of electron–phonon interaction will also produce a coupling between the electron spin and the strain field of the crystal lattice via the spin = orbit interaction. The above coupling is of fundamental origin. The consideration of spin–orbit interaction will lead to the electronic Hamiltonian density [5]:

$$H_{SO} = -\frac{i\mu_B}{4m}\left(\nabla\psi^+ .\sigma \times E\psi - \psi^+\sigma \times E.\nabla\times\right) \tag{5.21}$$

where μ_B = Bohr magnetor, and E = electric field.

Such effect is known as spin–orbit coupling. In order to preserve the gauge invariance of the theory, one has to replace the ordinary derivative ∇ by the covariant derivative D, producing

$$\widetilde{H_{SO}} = -\frac{i\mu_B}{4m}\left(\nabla\psi^+ .\sigma \times E\psi - \psi^+\sigma \times E.\nabla\psi\right)$$
$$+ i\frac{\mu_B}{4m} g\varepsilon_{ijk} E_j R_{kl}[\partial_l(\psi^+)\sigma_i \Psi - \psi^+\sigma_i\sigma_l\psi] \tag{5.22}$$

where repeated indices are to be summed over and g is the coupling constant defined previously. In the extra term appearing in (5.22), the electron spin is coupled to the space-like elastic field R_{kl} and to the electric field of the strained lattice.

In this treatment of gauge invariance approach to electron–phonon interaction, the phonons are related to a gauge field associated to local symmetry property as gauge bosons. This is in contrast to the general understanding of phonons as Goldstone bosons, associated with the spontaneous breaking of a global symmetry. In our treatment of the gauge invariance approach to electron–phonon interaction, we are dealing with local gauge invariance of non-Abelian group.

References

1. Gan, Woon Siong. 2007. *Gauge Invariance Approach to Acoustic Fields, Acoustical Imaging.* In vol. 29, ed. I. Akiyama, 389–394. Springer.
2. Yang, C.N., and R.L. Mills. 1954. Conservation of isotopic spin and isotopic gauge invariance. *Physical Review* 96 (1): 191–195.
3. Ryder, L.H. 1991. *Quantum Field Theory,* 2nd edn. Cambridge University Press.
4. Kittel, C. 1963. *Quantum Theory of Solids.* New York: John Wiley.
5. Dartora, C.A., and G.G. Cabrera. 2008. The electron–phonon interaction from fundamental local gauge symmetries in solids. Phys. Rev. B 78: 012403.

Chapter 6
B/A Nonlinear Parameter Acoustical Imaging

6.1 Introduction

The B/A nonlinearity parameter measures the nonlinearity of the equation of state for a fluid. It determines the distortion of a finite amplitude wave propagating in the fluid and the nonlinear correction to the velocity due to the nonlinear effects from the propagation of the finite amplitude wave. Besides this, it is also related to the molecular dynamics of the medium. Thus it can provide informations on the structural properties of the medium such as inter-molecular spacing, clustering, internal pressures, and the acoustic properties of the materials. Hence it is an important physical constant of characterizing the different acoustic materials and biological media [1, 2].

B/A nonlinear parameter is very useful in the medical applications of ultrasound, both in therapeutics and in diagnostics. In medical therapeutics, it enables the prediction of the temperature in the tissue during the ultrasonic hyperthermia treatment. In the diagnostics applications, B/A information can be used in the design and optimization of the ultrasound imaging devices.

There are two basic approaches in the experimental measurement of the B/A nonlinearity parameter: the thermodynamic method and the finite amplitude method.

6.2 The Thermodynamic Method

6.2.1 Theory

The equation of state $p = p(\rho, s)$ of a liquid, in thermodynamics can be expanded into a Taylor series along the isentrope $s = s_0$, as [3]

$$p - p_0 = A\left(\frac{\rho'}{\rho_0}\right) + \frac{B}{2!}\left(\frac{\rho'}{\rho_0}\right)^2 + \frac{C}{3!}\left(\frac{\rho'}{\rho_0}\right)^3 \tag{6.1}$$

© Springer Nature Singapore Pte Ltd. 2021
W. S. Gan, *Nonlinear Acoustical Imaging*,
https://doi.org/10.1007/978-981-16-7015-2_6

where $\rho' = \rho - \rho_0 =$ excess density, p = instantaneous pressure, $\rho =$ instantaneous density of the liquid disturbed by the ultrasonic wave propagation, p_0, $\rho_0 =$ their unperturbed (ambient) values and

$$A = \rho_0 \left(\frac{\partial p}{\partial \rho} \right)_s \Bigg]_{\rho=\rho_0} = \rho_0 c_0^2 \tag{6.2}$$

$$B = \rho_0^2 \left(\frac{\partial^2 p}{\partial \rho^2} \right)_s \Bigg]_{\rho=\rho_0} \tag{6.3}$$

$$C = \rho_0^3 \left(\frac{\partial^3}{\partial \rho^3} p \right)_s \Bigg]_{\rho=\rho_0} \tag{6.4}$$

where s = specific entropy, $c_0 =$ isentropic small signal sound speed. Subscript s denotes constant entropy process which is a condition for the ultrasound propagation. Also the partial derivatives are evaluated at the unperturbed state of (ρ_0, s_0).

From (6.2) and (6.3), B/A the nonlinearity parameter which is the ratio of the quadratic to the linear term in the Taylor series can be expressed as:

$$\frac{B}{A} = \frac{\rho_0}{c_0^2} \left(\frac{\partial^2 p}{\partial \rho^2} \right)_s \Bigg]_{\rho=\rho_0} \tag{6.5}$$

With the definition of sound speed $c^2 = \left(\frac{\partial p}{\partial \rho} \right)_s$, (6.5) can be written as

$$\frac{B}{A} = 2\rho_0 c_0 \left(\frac{\partial c}{\partial p} \right)_s \Bigg]_{\rho=\rho_0} \tag{6.6}$$

The partial derivative $\left(\frac{\partial c}{\partial p} \right)_s \Big]_{\rho=\rho_0}$ can be further expanded [3] giving

$$\frac{B}{A} = 2\rho_0 c_0 \left(\frac{\partial c}{\partial p} \right)_T \Bigg]_{\rho=\rho_0} + (2 c_0 \, Tq/c_0 c_p) \left(\frac{\partial c}{\partial T} \right)_p \Bigg]_{\rho=\rho_0} \tag{6.7}$$

where $\rho_0 =$ density of undisturbed medium, $c_0 =$ sound velocity for acoustic waves of infinitesimal amplitude, c = measured sound velocity at given temperature and pressure, p = acoustic pressure, T = absolute temperature in Kelvin, $c_p =$ specific heat at constant pressure, and q = isobaric volume coefficient of thermal expansion $= \left(\frac{1}{V} \right) \left(\frac{\partial V}{\partial T} \right)_p$.

6.2.2 Experiment

Equation (6.7) indicates that it is necessary to measure the sound velocity as a function of pressure and temperature accurately in the investigated medium in order to determine B/A. To use the thermodynamic method, a velocimeter is required. This is a vessel of known length L, comprising the test liquid and the transmitter–receiver equipment inserted in a liquid-filled pressure vessel such as water or oil that is in turn submerged in a bath with controlled temperature (Fig. 6.1).

The first term of Eq. (6.7) is known as the isothermal nonlinear parameter and the second term is referred to as the isobaric nonlinear parameter. This second term of Eq. (6.7) is used for the thermodynamic method for determination of B/A. The speed of sound measurement can be performed with different techniques to infer the travel time (time of flight), t_{tr} of the sound wave through the velocimeter of known length L. (Fig. 6.1). The first paper on the determination of B/A using the thermodynamic method is that of Beyer [3]. Greenspan and Tschiegg [5] used a sing-around circuit to infer t_{tr}, t_{tr} is inferred through the pulse repetition frequency (PRF) as the circuit allows for the triggering of the generator to send a pulse once the preceding pulse is received. The accuracy of the measurement was improved by Greenspan et al. [6] by adjusting the PRF of the generator for a new pulse to be transmitted when the echoes of the previously transmitted pulse were superimposed on the receiver.

In this way, the device can measure the speed of sound from the pulse transit time t_{tr}, inferred from the distance travelled by the pulse equal to twice the the length of the vessel and the PRF:

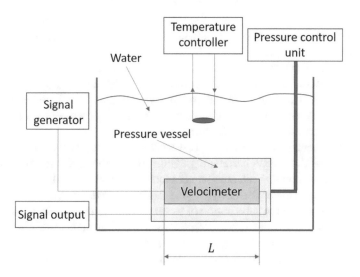

Fig. 6.1 Schematic simplified diagram of the typical setup used for a thermodynamic B/A measurement (After Panfilova [4])

$$C = 2L/t_{tr} \tag{6.8}$$

The accuracy of the thermodynamic method with taking into account the measurement uncertainties in temperature and pressure can result in a global uncertainty of the B/A estimation within 5%, for tissue and 3% for liquid. The higher uncertainty for tissue samples is due to the inhomogeneous speed of sound [7]. Also Law et al. [7] demonstrates the dependence of B/A on the structural hierarchy of biological chemical composition of biological material solution and the chemical composition of biological solutions.

6.3   The Finite Amplitude Method

The finite amplitude method for B/A measurement was introduced by Beyer [3] based on the Fubini [8] expansion to the nonlinear wave equation for a lossless media. Fubini expansion provided the distortion of acoustic waves measured by the strength of the second harmonics. Cobb [9] improved the results with the incorporation of attenuation and dissipation. Cobb [9]'s method has been widely used in the measurement of the nonlinearity parameter of biomaterials and biological tissues [10, 11]. In this method, the second harmonic pressure p_2 was given by [9]:

$$|p_2| = \left[1 + \frac{B}{2A}\right] \frac{(e^{-2\alpha_1 x} - e^{-\alpha_2 x})}{2\rho_0 c_0^3 (\alpha_2 - 2\alpha_1)} p_1^2 \tag{6.9}$$

where p_1 = first harmonic, α_1, α_2 = attenuation coefficients corresponding to p_1 and p_2 respectively.

In Eq. (6.9), $\beta = B/2A$ = nonlinearity coefficient. The first term "1" in Eq. (6.9) accounts for the self-convective effect contributing to the nonlinearity of propagation of acoustic waves arising from the continuity equation [11]. Together with B/A which is the inherent material nonlinearity parameter, it describes the distortion of propagating sound waves in the medium.

The experimental measurement of B/A can be performed by the measurement of the values of p_2 the second harmonic pressure and p_1, the fundamental pressure as well as the corresponding attenuation coefficients α_2 and α_1. at different locations away from the transducer source by varying x. Then the nonlinearity parameter B/A can be determined by the substitution of the experimental data in Eq. (6.9) [9].

The finite amplitude methods can be classified into three main categories: deriving B/A directly from the wave's shape, from the second harmonic component and from the fundamental component. A detailed description of each category of methods is given below.

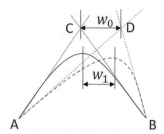

Fig. 6.2 Illustration of the waveform deformation. The image shows the positive half-cycles of an undistorted sinus wave (solid line), and of a distorted wave (dashed line), where the distance w_1 is the distance between the maxima of these waves, and w_0 is the distance between points C and D, which are the intersections of the tangents at points A and B (After Panfilova [4])

6.3.1 The Wave Shape Method

The earliest works with the observations of the wave shape to measure B/A were performed with optical methods [4] (Fig. 6.2). When an optical wave propagates in a direction perpendicular to the ultrasonic beam, the initially flat wavefront of the optical wave is modulated in phase according to the velocity profile of the ultrasonic wave, and distorted due to the nonlinear propagation. In this way, the measurement of the diffraction of light allows for the reconstruction of the ultrasonic wave's velocity profile and quantification of distortions by extracting w_0 or w_1 from its shape (Fig. 6.2). This leads to the derivation of B/A.

6.3.2 Second Harmonic Measuements

The amplitudes of the higher harmonics in the preshock region ($\sigma \leq 1$) of a plane wave in a lossless medium can be given by [13]:

$$p(\sigma) = \left[\frac{p(0)}{n\sigma} \right] J_n(n\sigma) \tag{6.10}$$

where n = position integer indicating the number of the harmonic: the fundamental $p(0)$ at the source and higher harmonics p, and σ = shock parameter = $2\pi f\, p_1$ $(0)Z\beta/\rho_0 c_0^3$. Equation (6.10) is the Fubini solution [12] of the lossless Burgers' equation given by

$$\frac{\partial p}{\partial x} - \frac{\beta}{\rho_0 c_0^3} p \frac{\partial p}{\partial \tau} = 0 \tag{6.11}$$

where τ = retarded time = t $-$ (z/c_0).

Expanding the Bessel function as a power series and neglecting higher order terms, one can write the amplitude of the 2nd harmonic as

$$p_2(z) = \left(\frac{B}{A} + 2\right)\frac{\pi f_z}{2\rho_0 c_0^3} p_1^2(0) \tag{6.12}$$

The above equation shows the quadratic dependence on the transmitted pressure amplitude $p_1(0)$ at the source and that the amplitude of the 2nd harmonic increases proportionally to B/A, to the frequency of the signal f and to the distance z from the source.

Dunn et al. [12] considered both attenuation and diffraction effects and obtained the following expression to the 2nd harmonic:

$$p_2(z) = \frac{\left(2 + \frac{B}{A}\right)\pi f z}{2\rho_0 c_0^3} p_1^2(0)\, e^{-\left(\alpha_1 + \frac{\alpha_2}{2}\right)z}\, F(z) \tag{6.13}$$

where F(z) = diffraction correction factor, α_1, α_2 = attenuation coefficients of the fundamental and its harmonic and f = frequency of the transmitted signal.

The above theory is the basis of the finite amplitude method. Numerical works have been performed for the estimation of B/A through 2nd harmonic measurement. In most of these works, the 2nd harmonic is measured in the nearfield of a plane piston source, enabling the plane wave approximation.

6.3.3 Measurement from the Fundamental Component

There is an increase in absorption with signal intensity. At first, this was attributed to cavitation [13]. However, later, it was realized to be the result of energy transfer from the fundamental to higher harmonics [14]. Moreover, nonlinear attenuation increases with nonlinear effects which grow with source amplitude. This limits the sound power that can be delivered to a certain depth.

Due to energy transfer to higher harmonics, in the shock-free or pre-shock region of $\sigma < 1$ of a lossless medium, the fundamental component of a plane wave will increase as follows:

$$p_1(z) = p_0\left(1 - \frac{1}{8}\sigma^2\right) = p_0\left(1 - \frac{1}{2}\left[\left(1 + \frac{1}{2}B/A\right)z\pi f_0 p_0/\rho_0 c_0^3\right]^2\right) \tag{6.14}$$

For the measurement of B/A in homogeneous media, two modes have been used: the transmission mode and the echo-mode. Hikata et al. [15] used the finite amplitude loss technique and performed the transmission mode measurement. Here the dependence of the received signal pressure p(z) on the intensity of the transmitted signal pressure p(0) was recorded at a fixed distance. They extracted β from the slope of the linear dependence by expressing the dependence of p(0)/p(z) via p(0). They discussed the trend of this finite amplitude loss technique to yield values higher than those acquired using the thermodynamic method and the light diffraction method. Although their method is limited to homogeneous media, it is rather simple, requiring only a pair of transducers with a similar resonance frequency. For the echo mode, Byra et al. [16] applied the echo-mode method using the lossless plane wave theory based on Eq. (6.14) to determine B/A of water. As in echo-mode imaging, the backscattered wave was assumed to travel linearly. They used the Verasonics research scanner equipped with a linear array probe L_{12-5} to image a set of reflecting wires positioned at different depths in water. They were able to observe a portion of the fundamental saturation and by fitting Eq. (6.14) to this curve to extract the water's β.

6.4 B/A Nonlinear Parameter Acoustical Imaging

6.4.1 Theory

The above section deals with the methods for the measurement of the B/A nonlinear parameter. In this section, the nonlinear acoustical imaging based on the B/A nonlinear parameter will be described based on the works of Varay et al. 17]. The equation describing the main effects of nonlinear ultrasound propagation usually is given by the KZK equation [18, 19]:

$$\frac{\partial^2}{\partial t^2}p = (\beta/2\rho_0 c_0^3)\frac{\partial^2}{\partial t^2}p^2$$
$$+ \frac{\alpha}{2c_0^3}\frac{\partial^3}{\partial t^3}p + \frac{c_0}{2}\left[\frac{\partial^2}{\partial x^2}p + \frac{\partial^2}{\partial y^2}p\right] \qquad (6.15)$$

where α = attenuation coefficient.

Here the first term on the right hand side of the equation describes the nonlinear effect, the second term represents the attenuation of tissue during the propagation and the last term provides the diffraction effect of the probe. If the diffraction effects are neglected as for the case of a plane wave, the above equation reduces to the Burgers equation.

A. The 2nd Harmonic Method

For a nonlinear medium with attenuation, the increase of the second harmonic p_2 along the propagation axis z can be given by [20]:

$$p_2(z) = \frac{^?\omega p_0^2}{2\rho_0 c_0^3}\exp(-\alpha_2 z)\int_0^z \beta(u)\exp[-(2\alpha_1 + \alpha_2)u]\,du \qquad (6.16)$$

For a theoretical medium without attenuation and with constant nonlinear parameter, (6.16) reduces to [21]:

$$p_2(z) = \frac{{}^?\omega p_0^2}{2\rho_0 c_0^3}\beta z \tag{6.17}$$

For nonconstant nonlinear parameter in attenuating media along the propagation axis, (6.16) can be rewritten as:

$$B(z) = (2\rho_0 c_0^3/\omega p_0^2)\left[\frac{dp_2(z)}{dz} + \alpha_2 p_2(z)\right]e^{2\alpha_1 z} \tag{6.18}$$

B.   The Comparative Method

The comparative method also known as the insertion substitution method is a common experimental method in B/A nonlinear parameter acoustical imaging. It was developed by Gong et al. [22]. To use this method in inhomogeneous medium, Eq. (6.16) has to be used. For the reference medium with a constant nonlinear parameter β_0, Eq. (6.16) is rewritten as:

$$p_{20}(z) = (\omega p_0^2/2\rho_0 c_0^3)\beta_0\left[\exp(-2\alpha_1 z) - \exp(-\alpha_2 z)\right]/(\alpha_2 - 2\alpha_1) \tag{6.19}$$

For a known medium (0) and an unknown medium (i) with constant nonlinear parameter, the ratio between second harmonic pressures is related to the nonlinear parameter as [23]:

$$\beta_1 = \left[\beta_0(\rho c^3)_i p_{2i}\right]/\left[(\rho c^3)_0 p_{20}\right] \tag{6.20}$$

In general, the ratio between the two second harmonics is given as:

$$p_{2i}(z)/p_{20}(z) = \left[(\rho c^3)_0/(\rho c^3)_i\right](\alpha_2 - 2\alpha_1)/\{\left[\exp(-2\alpha_1 + \alpha_2)z\right] - 1\}$$
$$\int_0^z \beta_i(u)\exp[(-2\alpha_1 + \alpha_2)u]/\beta_0\,du \tag{6.21}$$

β_i can then be isolated and deriverted to extract the nonlinear parameter from nonconsant nonlinear parameter medium:

$$\beta_i(z) = \left[\beta_0(\rho c^3)_i\right]/(\rho c^3)_0\left[\frac{p_{2i}}{p_{20}} + \frac{d}{dz}\left(\frac{p_{2i}}{p_{20}}\right)\{1 - \exp[(2\alpha_1 - \alpha_2)z]/(\alpha_2 - 2\alpha_1)\}\right] \tag{6.22}$$

Simulations have been done based on a KZK simulator [24].

6.4.2   Simulation

The simulated medium has a sound velocity of 1500 ms^{-1} and a density of 1000 kgm^{-3}. The B/A values of the medium evolve along the 150 mm of the propagation. The B/A parameter for the medium is defined as follows:

$$\frac{B}{A} = \begin{bmatrix} 3 & \text{for } 0\,\text{cm} \leq z \leq 4\,\text{cm} \\ 3 + 5(z-4) & \text{for } 4\,\text{cm} < z \leq 5\,\text{cm} \\ 8 & \text{for } 5\,\text{cm} < z \leq 9\,\text{cm} \\ 8 - 5(z-9) & \text{for } 9\,\text{cm} < z \leq 10\,\text{cm} \\ 3 & \text{for } 10\,\text{cm} < z \end{bmatrix}$$

The frequency of the transmitted ultrasound signal is set between 1 and 6 MHz and the ultrasound probe used in the simulator is based on the characteristics of a real linear array with 128 elements and transmit focus set at 71 mm. Two different cases were used for the simulation. The first was with the diffraction effect of the probe turned off and the transducer elements transmit an ideal wave which did not interact with neighbour's transmission. The second case was with the diffraction effect of the probe taken into consideration. Here the transducer elements transmit an ultrasound wave which was propagating in the entire space with interaction with the other elements' transmissions.

The simulator consists of the three characteristic terms of the KZK equation. The nonlinear effect, the attenuation of the tissue during the propagation and the diffraction effect of the probe can be set. The acoustic pressure profile can be obtained at each point of the z axis. From it, the fundamental or second harmonic components of the acoustic pressure can be extracted.

When the diffraction effects are set off, the KZK equation reduces to the Burgers equation. The maximum error on the B/A parameter computed from the comparative method along the propagation distance is only 2.5%.

With the diffraction effect of the probe set on, then the B/A parameter extracted according to the same formulae used above, the results are not as good as without diffraction. Here only the comparative method can be considered acceptable.

6.4.3   Experiment [17]

The experiment was based on the comparative method and the Ultrasonix SONIX RP scanner was used. An elastography phantom was used to simulate two different media, the interior and the exterior of the inclusion. The reference medium with constant B/A parameter has been defined on the left of the phantom in a homogeneous region. The second-order ultrasound field (SURF) technique was shown in Fig. 6.3. The comparative method at the first order was used. The latter does not depend on the

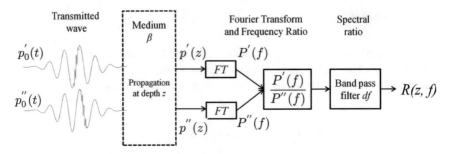

Fig. 6.3 SURF method: transmission of two consecutive waves with the different high-frequency pulse localizations. The nonlinear coefficient is computed from the pair of recorded pulses (After Varray [19])

probe focalization depth and was more homogeneous. Besides this, the investigation depth was increased.

6.4.4   Image Reconstruction with Computed Tomography

This is B/A nonlinear parameter acoustical images in a heterogeneous medium with through-transmission measurements and image reconstruction using computed tomography (CT). The through-transmission measurement is repeated for several sample rotation and translation configuration. The Radon transform is used in the reconstruction with the resulting image resolution determined by the number of employed rotation angles. The reflection mode measurement follows the same procedure except using the source as the receiver. The signal is reflected from a reflective plate on the opposite side of the sample after transmission through the tissue. The implemented CT allowed for the translation of the sources and receiver and rotation of the sample along the sample length. The receiver was positioned in the nearfield of the transmitting transducer. Gong et al. [25] extended their previous work [26] on through transmission mode to reflection mode. Here the tissue sample was positioned in water between the source and a reflected plate where the reflective plate replaced the receiver. An example of this experiment was shown in Fig. 6.4 which showed the image of a two-layered tissue structure with porcine liver surrounded by porcine fat, submerged in water.

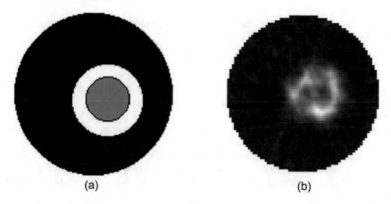

(a) (b)

Fig. 6.4 **a** Model of the imaged media: porcine liver surrounded with porcine fat, submerged in water. **b** The acquired reflection-mode tomographic image, utilizing the finite amplitude insert-substitution method (After Panfilova [4])

References

1. Law, W., et al. 1985. Determination of the nonlinearity parameter B/A of biological media. *Ultrasound in Medicine & Biology* 11 (2): 307–318.
2. Errabolu, R.L., et al. 1988. Measurement of ultrasonic nonlinear parameter in excised fat tissues. *Ultrasound in Medicine & Biology* 14 (2): 137–147.
3. Beyer, R.T. 1960. Parameter of nonlinearity in fluids. *The Journal of the Acoustical Society of America* 32: 719–721.
4. Panfilova, A., R.J.G. van Slovun, H. Wijkstra, O.A. Sapozhnikov, and M. Mischi. 2021. A review on B/A measurement methods with a clinical perspective. *The Journal of the Acoustical Society America* 149 (4): 2200–2237.
5. Greenspan, M., and C.E. Tschiegg. 1957. Speed of sound in water by a direct method. *Journal of Research of the National Bureau of Standards* 59 (4): 249–254.
6. Wilson, W.D. 1959. Speed of sound in distilled water as a function of temperature and pressure. *Journal of the Acoustical Society of America* 31: 1067–1072.
7. Law, W.K., L.A. Frizzell, and F. Dunn. 1985. Determination of the nonlinearity parameter B/A of biological media. *Ultrasound in Medicine and Biology* 11 (2): 307–318.
8. Fubini, G.E. 1935. Anomalie nella propagazione di onde acustiche di grande ampiezza. Industrie Grafiche Italiane Stucchi.
9. Cobb, W.N. 1983. Finiteamplitude method for the determination of the acoustic nonlinearity parameter B/A. The Journal of the Acoustical Society of America 73(5): 1525–1531.
10. Zhu, Z., et al. 1983. Determination of the acoustic nonlinearity parameter B/A from phase measurements. *Journal of the Acoustical Society of America* 74 (5): 1518–1621.
11. Sehgal, C., et al. 1986. Measurement and use of acoustic nonlinearity and sound speed to estimate composition of excised livers. *Ultrasound in Medicine & Biology* 12 (11): 865–874.
12. Hamilton, M.A., and D.T. Blackstock. 1988. On the coefficient of nonlinearity β in nonlinear acoustics. *Journal of the Acoustical Society of America* 83 (1): 74–77.
13. Dunn, F., W.K. Law, and L.A. Frizzell. 1982. Nonlinear ultrasonic propagation in biological media. *British Journal of Cancer Supplementary* 45: 55.
14. Fox, F.E., and G.D. Rock. 1941. Ultrasonic absorption in water. *Journal of the Acoustical Society of America* 12: 505–510.
15. Fox, F.E., and W.A. Wallace. 1954. Absorption of finite amplitude sound waves. *Journal of the Acoustical Society of America* 26: 994–1006.

16. Hikata, A., H. Kwun, and C. Elbaum. 1980. Finite amplitude wave propagation in solid and liquid He 4. *Physical Review B* 21 (9): 3932–3939.
17. Byra, M., J. Wójcik, and A. Nowicki. 2017. Ultrasound nonlinearity parameter assessment using plane wave imaging. In *Proceedings of the 2017 IEEE International Ultrasonics Symposium (IUS), Washongton, D.C.*, 1511–1516, September 6–9, 2017.
18. Varray, F., O. Basset, P. Tortoli, and C. Cachard. 2011. Extensions of nonlinear B/A parameter imaging methods for echo mode. *IEEE Transactions on Ultrasonics, Ferroelectgrics, and Frequency Control* 58 (6): 1232–1244.
19. Zabolotskaya, E., and R. Khokhlov. 1969. Quasi-plane waves in the nonlinear acoustics of confined beams. *Soviet Physics Acoustics* 15: 35–40.
20. Kuznesov, V. 1970. Equation of nonlinear acoustics. *Soviet Physics* 16: 749–768.
21. Zhang, D., and X. Gong. 1999. Experimental investigation of the acoustic nonlinearity parameter tomography for excised pathological biological tissue. *Ultrasound in Medicine & Biology* 25: 593–599.
22. Law, W.K., L.A. Frizzell, and F. Dunn, Ultrasonic determination of the nonlinearity parameter B/A for biological media. The Journal of the Acoustical Society of America 69: 1210–121.
23. Gong, X., R. Feng, C. Zhu, and T. Shi. 1984. Ultrasonic investigation of the nonlinearity parameter B/A in biological media. *Journal of the Acoustical Society of America* 76: 949–950.
24. Wallace, K.D., C.W. Lloyd, M.R. Holland, and J. Miller. 2007. Finite amplitude measurements of the nonlinear parameter B/A for liquid mixtures spanning a range relevant to tissue harmonic mode. *Ultasound in Medicine & Biology* 33: 620–629.
25. Lee, Y.S., and M.F. Hamilton. 1995. Time-domain modelling of pulsed finite-amplitude sound beam. *Journal of the Acoustical Society of America* 97: 906–917.
26. Gong, X., D. Zhang, J. Liu, H. Wang, Y. Yan, and X. Xu. 2004. Study of acoustic nonlinearity parameter imaging methods in reflection mode for biological tissues. *Journal of the Acoustical Society of America* 116 (3): 1819–1825.

Chapter 7
Ultrasound Harmonic Imaging

7.1 Theory of Ultrasound Harmonic Imaging

Here an introduction to the theory of nonlinear acoustics will be given. This will start with the derivation of a nonlinear acoustics wave equation. Only the one-dimensional case will be considered and the sound wave will be propagating in the x direction.

To consider the mass conservation, one considers a constant mass Δ_m between x and $x + \Delta_x$ with side surface S. Let the unidirectional state be defined by density ρ_0 and acoustic pressure p_0. Then the mass element Δ_m can be written as $\Delta_m = \rho_0 S \Delta_x$. When subjected to the disturbance moving through the mass element caused by the propagating sound wave, the same mass is now found between the new positions x $+ \xi$ and $x + \xi + \Delta_x + \Delta_\xi$ where ξ = particle displacement with

$$\Delta \xi = \left(\frac{\partial \xi}{\partial x} \right) \Delta x \tag{7.1}$$

One has

$$\Delta m = \rho S \left(1 + \frac{\partial \xi}{\partial x} \right) \Delta x \tag{7.2}$$

From (7.1) and (7.2), the following can be obtained:

$$\rho_0 = \rho \left(1 + \frac{\partial \xi}{\partial x} \right) \tag{7.3}$$

Next is to derive the equation of motion for the same mass element Δ_m. Apply Newton's second law:

$$F = ma = \Delta m \frac{\partial^2}{\partial t^2} \xi \tag{7.4}$$

© Springer Nature Singapore Pte Ltd. 2021
W. S. Gan, *Nonlinear Acoustical Imaging*,
https://doi.org/10.1007/978-981-16-7015-2_7

With no external forces, and with Eq. (7.2),

$$p(x)S - p(x + \Delta x)S = \Delta x S \rho_0 \frac{\partial^2}{\partial t^2} \xi \tag{7.5}$$

Expanding the pressure in a series $p(x + \Delta x) = p(x) + \frac{\partial p}{\partial x} \Delta x$ and inserting in (7.5):

$$\frac{\partial p}{\partial x} = \rho_0 \frac{\partial^2}{\partial t^2} \xi \tag{7.6}$$

We first derive the nonlinear acoustic wave equation for a gas [1]. The density is given by $\rho = m/V$ or $V = m/\rho$, where $V =$ volume. Consider the small mass element Δm and for isentropic gases, one has the adiabatic gas law:

$$p\left(\frac{\Delta m}{\rho}\right)^\Upsilon = \text{constant} = p_0 \left(\frac{\Delta m}{\rho_0}\right)^\Upsilon \tag{7.7}$$

where $\gamma = c_p/c_V =$ specific heat.

From (7.7), m can be eliminated,

$$p = p_0 \left(\frac{\rho}{\rho_0}\right)^\Upsilon \tag{7.8}$$

Rewriting (7.3) as $\rho = \rho_0/(1 + \frac{\partial \xi}{\partial x})$ and inserting into (7.8), $p = p_0/\left(1 + \frac{\partial \xi}{\partial x}\right)^\Upsilon$ is obtained.

Inserting this into (7.6), resulting in

$$\rho_0 \frac{\partial^2}{\partial t^2} \xi = -\left(\frac{\partial}{\partial x}\right) \frac{p_0}{\left(1 + \frac{\partial \xi}{\partial x}\right)^\Upsilon} \tag{7.9}$$

which can be arranged as

$$\frac{\partial^2}{\partial t^2} \xi = \Upsilon \left(\frac{p_0}{\rho_0}\right) \frac{\frac{\partial^2 \xi}{\partial x^2}}{\left(1 + \frac{\partial \xi}{\partial x}\right)^{\Upsilon+1}} \tag{7.10}$$

Here $c_0^2 = \Upsilon \frac{p_0}{\rho_0}$ is found to be the speed of sound and $(\Upsilon + 1)/2 = \epsilon =$ nonlinear parameter.

Then Eq. (7.10) can be written as:

$$\frac{\partial^2}{\partial t^2}\xi = c_0^2 \frac{\frac{\partial^2}{\partial x^2}\xi}{\left(1 + \frac{\partial \xi}{\partial x}\right)^{2\epsilon}} \tag{7.11}$$

This equation is known as the Earnshaw equation derived in 1860 [2].

To study quadratic nonlinearity or higher orders of nonlinearity, $1/\left(1 + \frac{\partial \xi}{\partial x}\right)^{2\epsilon}$ can be expanded in a Fourier series as,

$$1/\left(1 + \frac{\partial \xi}{\partial x}\right)^{2\epsilon} = 1 - 2\epsilon \frac{\partial \xi}{\partial x} + \frac{2\epsilon(\epsilon - 1)}{2}\left(\frac{\partial \xi}{\partial x}\right)^2 + \dots \tag{7.12}$$

Inserting Eq. (7.12) into Eq. (7.11), one has

$$\frac{\partial^2}{\partial t^2}\xi - c_0^2 \frac{\partial^2}{\partial x^2}\xi = -2\epsilon c_0^2 \frac{\partial \xi}{\partial x}\frac{\partial^2}{\partial x^2}\xi + \epsilon(2\epsilon - 1)c_0^2(\frac{\partial \xi}{\partial x})^2\frac{\partial^2}{\partial x^2}\xi + \dots \tag{7.13}$$

The first term on the right hand side represents the quadratic nonlinearity and the second term represents the cubic nonlinearity. Equation (7.13) can be written in the following form for a more general description, valid for any medium:

$$\underbrace{\frac{\partial^2}{\partial t^2}\xi}_{\text{I}} - \underbrace{c_0^2 \frac{\partial^2}{\partial x^2}\xi}_{\text{II}} = \underbrace{c_0^2 \frac{\partial^2}{\partial x^2}(\beta \xi^2 + \delta \xi^3)}_{\text{III}} \tag{7.14}$$

The parameters β and δ can be found from experiments. In (7.14), I stands for the linear term, II for the quadratic nonlinearity and III for the cubic nonlinearity.

As an illustration of nonlinear phenomenon, take a sinusoidal wave, the initial shape is that of a sinusoid but as the wave propagates through a nonlinear medium, the shape becomes like an N-wave. Here the sound wave will be expressed by the particle velocity $u = \partial \xi/\partial t$, instead of the particle displacement ξ. In order to enable the observation of changes in the shape, the retarded time τ is introduced for a wave propagating in the x direction:

$$\tau = t - x/c_0 \tag{7.15}$$

The nonlinear phenomenon is shown by the higher magnitude of particle velocity u due to higher acoustic pressure p and higher density ρ and subsequently the sound speed c is increased. The sound wave moves to the left relative lower amplitudes and arrives earlier, with lower magnitude of particle velocity, the sound velocity is decreased. Shock waves will be produced. The distance x_{shock} is the distance where a shock appears, which is where the wavefront for the first is vertical at some point. This phenomenon can be described by the following equation [3]:

$$\frac{\partial u}{\partial x} = \frac{\epsilon\, u}{c_0^2} \frac{\partial u}{\partial \tau} \tag{7.16}$$

The equation has the following solution:

$$u = u_0(\tau + \frac{\epsilon}{c_0^2} ux) \tag{7.17}$$

Equation (7.16) is known as the Riemann equation. It is the simplest nonlinear equation in acoustics, showing only quadratic nonlinearity.

Taking the sound wave at the source to be represented by

$$u(x = 0, \tau) = u_0(\tau) = a_L \sin(\Omega_0\tau) + a_H \sin(\omega_0\tau) \tag{7.18}$$

with $\omega_0 \gg \Omega_0$.

The nonlinear interaction between these two waves can be obtained by investigating the solution (7.17) for a small step x away from the source (z also small). Inserting the signal (7.18) into the solution (7.17) will produce (with $z = \epsilon\, x/c_0^2$):

$$u(z, \tau) = a_L \sin(\Omega_0\tau[\tau + z(a_L \sin(\Omega_0\tau) + a_H \sin(\omega_0\tau))]$$
$$+ a_H \sin(\Omega_0[\tau + z(a_L \sin(\Omega_0\tau) + a_H \sin(\omega_0\tau))]) \tag{7.19}$$

Using the MacLaurin expansion in z:
$f(z) = f(0) + f(0)z + \dots$, one has

$$u(z, \tau) = \underbrace{a_L \sin(\Omega_0\tau) + a_L \cos(\Omega_0\tau)\Omega_0.a_L \sin(\Omega_0\tau)z}_{\text{I}}$$

$$\underbrace{+a_L \cos(\Omega_0\tau)\Omega_0 a_H . \sin(\omega_0\tau)z + a_H \sin(\omega_0\tau)}_{\text{II}}$$

$$\underbrace{+a_H \cos(\omega_0\tau)\omega_0 a_L \sin(\Omega_0\tau)}_{\text{III}} + \underbrace{a_H \cos(\omega_0\tau)\omega_0 a_H \sin(\omega_0\tau)z}_{\text{IV}} \tag{7.20}$$

The first and the fourth terms are the linear source frequency terms. The more interesting ones are the nonlinear interaction terms I, II, III, IV which are the second, third, fifth and sixth terms respectively. The trigonometric products in terms I can be simplified as

$$\cos(\Omega_0\tau) \sin(\Omega_0\tau) = \sin(2\Omega_0\tau)/2 \tag{7.21}$$

and the product terms in term IV as

$$\cos(\omega_0\tau) \sin(\omega_0\tau) = \sin(2\omega_0\tau)/2 \tag{7.22}$$

Both these products result in the creation of the first higher harmonic of the source frequencies.

The product with mixed frequencies in term II is

$$\cos(\Omega_0 \tau) \sin(\omega_0 \tau) = [\sin((\omega_0 + \Omega_0)\tau + \sin(\omega_0 - \Omega_0)\tau)]/2 \qquad (7.23)$$

There is an asymmetry between the upper and lower sidebands at frequency $\omega_0 + \Omega_0$ and $\omega_0 - \Omega_0$ because both terms II and III contribute. The combination of the two sideband terms II and III from Eq. (7.20) yield

$$u_{SB} = 2a_L a_H Z\{\Omega_0[\sin((\omega_0 + \Omega_0)\tau) + \sin((\omega_0 - \Omega_0)\tau)]/2$$
$$+ \{\omega_0[\sin(\omega_0 + \Omega_0)\tau) - \sin((\omega_0 + \Omega_0)\tau)]\}/2 \qquad (7.24)$$

$$= a_L a_{Hz}\{(\omega_0 + \Omega)\sin((\omega_0 + \Omega_0)\tau) - (\omega_0 - \Omega)\sin[(\omega_0 - \Omega_0)\tau)]\} \qquad (7.25)$$

There is an amplitude difference between the sidebands. The sideband with lower frequency is smaller than the upper by the factor $(\omega_0 - \Omega)/(\omega_0 + \Omega)$. Equation (7.25) shows how the sidebands around the high frequency are created, and that they are proportional to the nonlinear parameter ϵ, because z $=\epsilon$ x/c_0^2. With values used in measurements in experiments in inserting into the Eq. (7.20): $\omega = 15{,}000$, $\Omega = 2000$, $c_0 = 5500$, $a_L = a_H = 1$, x $= 0.1$ and $\epsilon = 500$, the spectrum can be plotted and shown in Fig. 7.1

The nonlinear wave modulation can be utilized for determination of the nonlinearity parameter G in a signal from the amplitude of the sidebands.

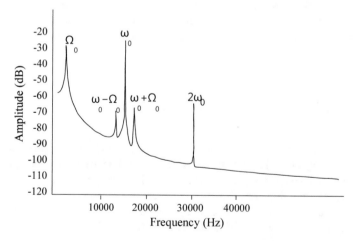

Fig. 7.1 Frequency response spectrum after nonlinear propagation the distance $x = 0.1$ m of two initial frequencies $\omega = 15{,}000$ Hz and $\Omega = 2000$ Hz (After Haller [4])

7.2   Methods Used to Isolate the Second Harmonic Signal Component

The usual method used to isolate the second harmonic signal component is the harmonic band filtering technique. The intensity of the second harmonic echo is 10–20 dB lower than that of the fundamental frequency. Hence one needs a low-noise, wide dynamic range receiver to receive and process the relatively weak signal [5].

An object within the ultrasonic field where the second harmonics is present will be interrogated by both the fundamental frequency and the harmonic frequency. Harmonics are generated as a result of propagation through tissue. This mechanism of second harmonics generation is different from that of harmonics due to the scattering from microbubble contrast agents, due to the expansion and contraction of the bubbles caused by the ultrasonic force. The detected signals or the returning echo will contain both the fundamental frequency component and the harmonic frequency component. After the reflection, the propagation of echoes at relatively low intensity will not generate additional harmonic components. A high-pass filter is used to remove the fundamental echo signal and only the tissue harmonic component is processed for image formation.

Ideally, there should not be overlapping between the fundamental and the harmonic bands. The transmitted pulse must be carefully shaped and controlled to prevent high-frequency components within the harmonic region. The presence of these high frequencies will give rise to echoes of the same frequency including clutter and noise. The fundamental and the harmonic bands can be separated by elongating the transmission pulse to form a narrow transmission bandwidth. This may degrade the axial resolution. However, the longer transmitted spatial pulse can be compensated to some extent by the increased detection frequency of the second harmonics.

7.3   Advantages of Harmonic Imaging

The unwanted effects of side lobes, grating lobes, and clutter are produced at the fundamental frequency. The echoes generated by the sources are also occurred at the fundamental frequency. They are suppressed in the harmonic image.

In harmonic imaging, reverberation artifacts are reduced. This is because the echoes from reflections in general do not generate harmonics due to their low amplitude. Multiple scattering and distortion at the fundamental frequency are suppressed in the harmonic image [6].

In harmonic imaging, there is reduction of acoustic noise. This enhances contrast resolution and border delineation. Hence the detection and characterization of low-contrast solid lesions are improved. Also acoustic enhancement and shadowing are more easily demonstrated [7].

The source of harmonic production is due to the high acoustic pressure occurs near the main beam axis and within the focal zone. The width of the harmonic beam is effectively narrower than the main beam at the fundamental frequency. The narrowing of beam width improves lateral resolutions [6].

7.4 Disadvantages of Harmonic Imaging

Due to the fact that only a small fraction of the transmitted probe energy is converted into the second harmonics and thus available for echo generation, the signal-to-noise ratio is lower and consequently harmonic imaging is less sensitive than conventional B-mode echocardiography.

The signal-to-noise ratio can be improved by lowering the transmitted frequency and reducing the receiver bandwidth. Due to the higher attenuation rates for harmonics, the overall penetration in harmonic imaging is less than that for conventional B-mode echocardiography. The drawback is not that severe because harmonics travel only a portion of the total path length.

Axial resolution for all reflections and contrast resolution for targets with small axial dimensions are likely to be inferior in harmonic imaging.

Due to the lack of harmonics in the incident ultrasound wave, structures located near the transducer are depicted with poor contrast.

7.5 Experimental Techniques in Nonlinear Acoustics

The simplest experimental method is known in the generation of higher harmonics. This is using the terms I with frequency Ω_0 and, IV with frequency ω_0 of Eq. (7.19):

$$\frac{1}{2}a_L^2 z \sin(2\Omega_0 \tau) \text{ and } \frac{1}{2}a_H^2 z \sin(2\Omega_0 \tau)$$

First the amplitudes of generated higher harmonics is investigated. A single frequency acoustic wave is introduced into an object by a piezoelectric transducer made of lead-zirconate titanate (PZT). As the sound wave is propagating in the nonlinear medium, it is distorted with the generation of nonlinear effect. The sound wave is then received by another PZT transducer and recorded on an oscilloscope where the frequency content can be presented. This accumulated effect is shown in Fig. 7.2a [4].

Here a pure sinusoidal wave is sent into materials with uniformly distributed nonlinearity at $x = 0$. Then the shape of the wave is investigated at discrete distances. The wave has become mildly distorted at distance x_a. This distorted wave is the input from distance x_a to x_b. The wave has become more distorted at distance x_b. After analysing the frequency contents of the these waves, it shows that the amplitude of

Fig. 7.2 Schematic wave propagating being distorted in a nonlinear medium. **b** Frequency content of response from an undamaged material. **c** Frequency content of response from a damaged material. (After Haller [4])

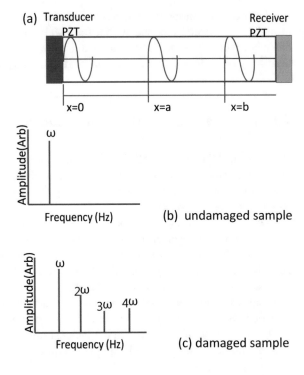

the frequencies 2ω, 3ω, …, $n\omega$ have decreased and this can be used as an indicator for nonlinearity and for damage detection. (Please refer to the difference between Fig. 7.2b and c). The nonlinearity factor ϵ can be calculated from this [8, 9].

7.6 Application of Ultrasound Harmonic Imaging to Tissue Imaging

Tissue harmonic imaging was introduced in 1997 [10] during the studies of the harmonic frequencies from insolating microbubbles realized that human tissues could also produce diagnostically useful harmonic waves. When the acoustic pressure in a soft tissue is small, that is less than 0.5 MPa, the tissues behave in a linear fashion and sound wave containing new frequencies are not generated. Nonlinear propagation occurs when high pressure sound wave, that is higher than 0.5 MPa propagates through a compressible medium and the transmitted ultrasound wave induces less compression than rarefaction. The production of harmonic sound waves including those from the edges of the sound beam is proportional to the square of the fundamental intensity. This is highest at the beam centre where the beam intensity is greatest and at the focal zone, where the beam is narrowest. Weaker sound waves produce little or no harmonics. These waves are from the edges of the sound beam and

Fig. 7.3 Comparison of tissue harmonic imaging (THI) with fundamental imaging. Sonograms of the kidney obtained at 4 MHz: **a** harmonic and **b** fundamental. (After Hedrick and Metzger [11])

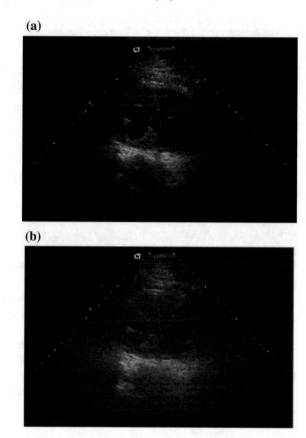

scattered echoes, side lobes and grating lobes. Images produced with tissue harmonic imaging are superior to conventional gray-scale images of cysts and abnormality that contain fat, calcium or air. There are several clinical advantages. A comparison of tissue harmonic imaging with fundamental imaging is shown in Fig. 7.3 [11].

7.7 Applications of Ultrasonic Harmonic Imaging to Nondestructive Testing

The experiment is performed using commercially available piezoceramic transducers (PZT) glued with epoxy. One PZT transducer acts as an actuator to generate the sound wave in the specimen and the other PZT transducer acts as a receiver of the sound wave. For the higher harmonics experiment, a single sinusoidal sound wave is sent to the specimen by the signal generator through one PZT transducer. The signal is distorted by the cracks and other defects on its way to the other PZT transducer acting

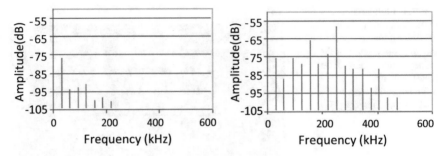

Fig. 7.4 Higher harmonics results of an undamaged circuit board ceramics on the left, and a damaged one on the right. (After Haller [4])

as a receiver. In the recorded frequency spectrum, higher harmonics are detected to indicate the amount of cracks or defects.

The sample used for the experiment is the ceramic semiconductor mounted to the printed circuit board. Higher harmonics was present.

When the semiconductor was damaged. This indicates that either the board itself gives nonlinear response or some component mounted was damaged. After testing a larger number of boards indicated that the nonlinear response arose from the board. When the semiconductors were damaged, the number of multiple harmonics increased and also the amplitude of the harmonics increased. This was shown in Fig. 7.4.

The method of harmonic imaging is convenient for circuit boards with no impacts is present because one has to be careful with the Semiconductor component.

7.8   Application of Ultrasound Harmonic Imaging to Underwater Acoustics

Ultrasound harmonic imaging can be applied in active sonar [12]. Prieur et al. [12] developed an active sonar that can receive at both the fundamental and second harmonic frequencies as an aid for target classification. They confirmed the presence of harmonic signals by measuring the pressure field radiated by the circular transducers with a centre frequency of 120 KHz for the first one and 200 KHz for the second one in a water tank up to 12 m range. These measurements also gave them the opportunity to compare with their numerical simulations. They then showed that second harmonic imaging could be used for target detection by imaging spherical targets using a pulse-echo technique. This showed better resolution capabilities compared to images obtained with the fundamental signal. They then used numerical simulations of the pressure field and the active sonar equation to estimate the maximum useful range of the fundamental signal. They also suggested the advantages of combining second harmonic signal with the fundamental signal.

After the confirmation experimentally from the measured profiles that higher harmonics were present and detectable, they set up an experiment for the second.

Experiment for the second harmonic pulse-echo imaging. They set up an experiment using a ES120-7C transducer with a centre frequency of 121 KHz to send a pulse that reflected on targets with the second harmonic at 242 KHz recorded by the ES200-7C transducer.

Both transducers were set side by side. The targets were four spheres. These spheres were made of tungsten carbide and the fourth sphere made of copper. They were all positioned on the horizontal plane which contained the propagation axis of both transducers. The transducer ES200-7C was connected directly to the oscilloscope. The transducers were rotated counter clockwise to cover an angular range of approximately 1.5–4.5°. 0° is the direction parallel to the wall of the water tank and the positive angle is taken in the clockwise direction. The recorded data were processed to filter out the pulse around the fundamental and the second harmonic bands.

The echoes from the three biggest spheres were clearly visible. The echo from the smallest sphere is barely noticeable when filtering around the fundamental frequency while filtering around the second harmonic it is clear. This is caused by the wider main lobe of the fundamental signal compared to the main lobe of the second harmonic signal.

The experiment shows that it is possible to use the second harmonic signal for imaging spheres. The image obtained by using the signal around the second harmonic frequency shows better resolving capabilities and reveals one target that fundamental imaging does not detect. The second harmonic signal has a larger main-lobe-to-side-lobe ratio and this is beneficial for target imaging. In a shallow-water environment, surface and bottom refractions of the side-lobes will cause perturbation to the sonar scanning at low grazing angles. There will also be perturbations from scatters situated in the propagation direction of the side lobes. Due to the lower side-lobe levels, the amplitude of these perturbations will be reduced in second harmonic imaging.

The combination of echoes around the fundamental and second harmonic frequency bands has the advantage that it gives an update that is twice the rate of a sonar receiving echoes around the fundamental frequency only. With the two images obtained, one can combine the high resolution of the second harmonic signal at short range and the long-range capability of the fundamental with a lower resolution.

The second harmonic signal exhibits low side-lobes relative to the main lobe. This is useful in many applications of sonar imaging. Combining echoes from the fundamental and second harmonic signals doubles the data rate per ping. In addition, the echoes at two different frequencies can contribute to target classification, for instance, living organisms, by comparing their frequency response. This is an application of the second harmonic signal in fisheries research.

High transmitted power is needed in this work on the application of second harmonic imaging to active sonar. With low input power the higher harmonics signals generated due to nonlinear propagation are negligible. Medium-to-high input powers are necessary. In the experiment of Prieur et al. [12], 1 kW input power was sufficient to achieve second harmonic imaging. By increasing input power there are limitations caused by cavitation, hard shock or saturation that all dissipate energy into the medium. Besides this, the receiver has to be sensitive enough to detect the low level of the echoes and the uncertainty of the recorded level has to be small for use in organism characterization.

References

1. Beyer, R.T. 1997. *Nonlinear acoustics.* Acoustical Society of America.
2. Earnshaw, S. 1860. On the mathematical theory of sound. *Philosophical Transactions of the Royal Society of London* 150: 133–148.
3. Enflo, B.D., and Hedberg, C.M. 2002. *Theory of nonlinear acoustics in fluids.* Kluwer Academic Publishers.
4. Haller, K. 2007. Nonlinear acoustics applied to nondestructive testing, Ph.D. thesis, Sweden: published by Blekinge Institute of Technology.
5. Hedrick, W.R., Hykes, D.L., Starchman, D.E. 2005. *Ultrasound Physics and Instrumentation.* 14th edition. MV, Mosby: St. Louis.
6. Lencioni, R., D. Cioni, and C. Bartolozzi. 2002. Tissue harmonic and contrast-specific imaging: back to gray scale in ultrasound. *European Radiology* 12: 151–165.
7. Tranquart, F., N. Grenier, V. Eder, and L. Pourcelot. 1999. Clinical use of ultrasound tissue harmonic imaging. *Ultrasound in Medicine and Biology* 25: 889–894.
8. Buck, O., W.L. Morris, and J.M. Richjardson. 1978. Acoustic harmonic generation at unbonded interfaces and fatigue cracks. *Applied Physics Letters* 35 (5): 31–373.
9. Barnard, D.J., G.E. Dace, D.K. Rehbein, and O. Buck. 1997. Acoustic harmonic generation at diffusion bond. *Journal of Nondestructive Evaluation* 16 (2): 77–89.
10. Averkiou, M.A., M.A. Roundhill, and D.R. Powers. 1997. A new imaging technique based on the nonlinear properties of tissues. *Proceeding IEEE Ultrasonic Symposium* 2: 1561–1566.
11. Hedrick, W.R., and L. Metzger. 2005. Tissue harmonic imaging, a review. *Journal of Diagnostic Medical Sonography* 21 (3): 183–189.
12. Prieur, F., Nasholm, S.P., Austeng, A., Tichy, F., and Holm, S. 2012. Feasibility of second harmonic imaging in active sonar: measurements and simulation. *IEEE Journal of Ocean Engineering* 37(3): 3467–3477.

Chapter 8
Application of Chaos Theory to Acoustical Imaging

8.1 Nonlinear Problem Encountered in Diffraction Tomography

Woon Siong Gan [1] is the first to introduce the concept of chaos theory to acoustical imaging in 1990. It was applied to the nonlinear problems encountered to obtain the expression for the scattered field in the imaging modality of diffraction tomography. Diffraction tomography is a common method used in medical ultrasound imaging such as for breast imaging. Breast tissue being inhomogeneous will cause, multiple scatterings on the sound wave propagating in the medium. This is a breakthrough in diffraction tomography as the usual approach is Born and Rytov approximations in the solving the nonlinear problem of the Lippman–Schwinger integral in diffraction tomography. These are linear and first order approximations.

Based on the mathematical theorem that every nonlinear differential equation has a chaos solution subjected to certain conditions. Since the Lippman–Schwinger integral is a nonlinear differential equation, it will have chaos solutions. Chaos phenomena have fractal characteristics geometrically. That is the chaos solution will give to fractal images. Since there is a key property of chaos that it is very sensitive to the initial condition, the fractal images produced should be able to show very fine changes in the condition of the object. For the case of breast imaging it will be useful for the detection of early stage cancer.

A background on chaos theory and fractal images will be given followed by the implementation in the solution of the Lippmann–Schwinger integral and then the formation of the fractal images. Then the application to medical ultrasound imaging will be given with an illustration by breast imaging.

© Springer Nature Singapore Pte Ltd. 2021
W. S. Gan, *Nonlinear Acoustical Imaging*,
https://doi.org/10.1007/978-981-16-7015-2_8

8.2 Definition and History of Chaos

The usual meaning of the word 'chaos' is a state of disorder or confusion and is equivalent to randomness. This can be applied to most cases but is not applicable when one considers the mathematical theory of chaos. The mathematical theory of chaos was discovered in the 1960s by Edward Lorenz, a meteorologist. He was attempting to predict the weather through the use of a mathematical model. He managed to generate a theoretical sequence of weather predictions using computer [2].

In a different attempt to observe the sequence of weather pattern, in order to save time, he ran the problem from the middle of the sequence. He was surprised to find that the results deviated from the previous sequence. He found that the error was due to a slight truncation in the decimal places of the values used. With this discovery, Lorenz began to establish the mathematical theory of chaos by relating this to the effect of a butterfly in the prediction of weather [2]. The flapping of a single butterfly's wing today produces a tiny change in the state of the atmosphere. Over a period of time, the atmosphere diverges from what it would have done. Therefore in one month time, a tornado that would have devastated the Indonesian coast does not happen. Or, perhaps the event that was not going to happen, happens [3]. These phenomena are known in chaos theory to be the sensitive dependence on initial condition.

Lorenz concluded that it is impossible to predict the weather since slight factor can affect it so drastically. He thereby proceeded to simplify his model for weather forecast and developed a set of three simple equations that have a sensitive dependence on initial conditions. These equations would generate a sequence of random behaviour. He was surprised to note that an infinite, double spiral known as the Lorenz Attractor was generated when the equations were plotted (Fig. 8.1).

The output fluctuated within the spirals. Gleick [4] discovered that these equations described the water wheel system exactly.

8.3 Definition of Fractal

The word 'fractal' was coined by Benoit Mandelbrot from the Latin adjective fractus. It has the meaning of being fragmented and irregular. It was created to describe a rough geometric figure that has the property of self-similarity or its statistical average. Self-similarity of a single item means that the item has a repetitive nature of itself when it is viewed under different length scales. This is shown in Fig. 8.2 which shows a Mandelbrot set, a computer-generated fractal. Figure 8.2a shows the Mandelbrot set at a particular viewing angle. By zooming in on this figure, a similar figure of the Mandelbrot set is obtained as shown in Fig. 8.2b (defined by the grey box). A further zooming reconfirms the repetitive nature of the fractal as shown in Fig. 8.2c.

Fig. 8.1   3-D Lorentz
Attractor. From http://astron
omy.swin.edu.au/pbourke/fra
ctals/lorenz/

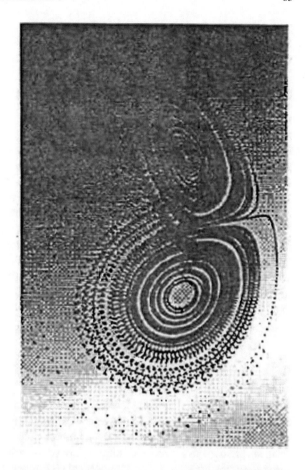

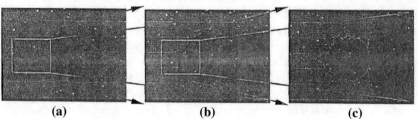

Fig. 8.2   Zooming in of Mandelbrot set (from left to right). From Hsiang (2002)

8.4   The Link Between Chaos and Fractals

The above is a general introduction to chaos and fractals. Next one will describe
the link between them. From careful observation, one will notice that fractals are
actually the visible representations of chaotic systems. If one examines the Lorenz

Table 8.1 Different classification of cellular automata [5]

Class 1	Evolution to a homogeneous state (an attractor)
Class 2	Evolution to isolated periodic segments
Class 3	Evolution that is always chaotic
Class 4	Evolution to isolated chaotic segments

attractor (Fig. 8.1) carefully one will notice that the figure is actually a fractal, since it is a visible representation of the chaotic system of the water wheel. An alternative way to study the link between chaos and fractals is to study cellular automata. This is a form of mathematical abstraction from a dynamical system. This has been used for the modelling of massively parallel computer and biological cell behaviour [5].

Cellular automata consists of a space of unit cells and these are initialized by '1' for living cell and '0' for unoccupied or dead cell. Besides this, the evolution of these cells is governed by a rule which bases the contents of the cell at time t by its contents at time $t - 1$. With this rule, different cellular automata can evolve which eventually end up with stable, fractal-like formation. With this classification rule for the cellular automata, a relationship given in Table 8.1 can be established with chaos [5].

8.5   The Fractal Nature of Breast Cancer

The fractal patterns inside cells can reveal breast cancer as this have been successfully shown by scientists at Mount Sinai School of Medicine (Fig. 8.3).

Traditionally pathologists must detect breast cancer by studying individual cells from suspicious tissues and checking for abnormal-looking cell shapes and features through subjective means. The Mount Sinai researchers have studied the distribution of chromatin and DNA–protein compounds which contain the chromosomes in a cell by looking within the cell nucleus using the analysis of images of actual breast cells. Like many other biological structures in nature, chromatin forms fractal patterns meaning that the chromatin has similar arrangement over a range of size scales. They apply techniques to study cells from 41 patients of whom 22 were known to have breast cancer from independent means. In a blind study, they correctly diagnosed 39 out of 41 cases, that is with a success rate of 95.1%. This was done by measuring the differences in lacunarity and differences in fractal dimensions between malignant and benign cells. Lucunarity is the size of gaps between chromatin regions in the nucleus and fractal dimension describes how fully a fractal object fills up the space is occupied.

There are other works [6–8] that confirmed the scattering of ultrasound to be chaotic in nature. With these works, one can confidently conclude that the fractal growth model can be used to represent the scattered field within the breast due to the fact that all chaotic systems can be represented by fractals.

Fig. 8.3 Surface plot of a
malignant breast epithelial
cell. From http://www.nyt
imes.com/specials/women/
warchive/970401_1732.htnl

Before going into details in the use of fractal growth model in diffraction tomography, some basic foundations have to be laid for the deep understanding of the theory and the mathematics behind the model.

8.6   Types of Fractals

There are two main categories of fractals: nonrandom fractals and random fractals.

8.6.1   Nonrandom Fractals

Nonrandom fractals [9] are man made and generated by a computer algorithm. To have a better understanding of this, the Sierpinski Gasket, a much studied nonrandom fractal will be discussed. The Sierpinski Gasket is defined operatively as an aggregation process that can be obtained using a simple iterative process.

The basic unit of the Sierpinski Gasket is a triangle (Fig. 8.4a) with a unit mass ($M = 1$) and a unit edge length ($L = 1$). In the first stage of iteration, three basic triangles are joined together to form Fig. 8.5b with mass of 3 and an edge length of 2.

With the density of the object defined as

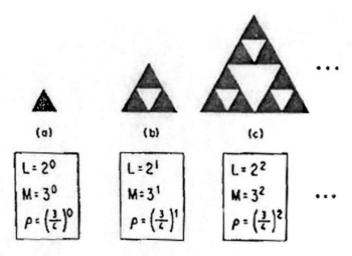

Fig. 8.4.   First few iteration stages in forming the Sierspinski Gasket. After Stanley [9]

Fig. 8.5 A log ρ (density) versus log L (length) plot for Sierpinski Gasket. After Stanley [9]

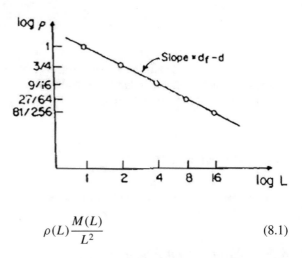

$$\rho(L)\frac{M(L)}{L^2} \tag{8.1}$$

implies that the density of the object actually decreases from unity to ¾. A further step of repeating the iterations, will reduce the density to $\left(\frac{3}{4}\right)^2$ (Fig. 8.5c). By plotting ρ against L on a double logarithmic graph paper, two striking features will be shown (Fig. 8.5):

1. $P(L)$ decreases monotonically with L, that is without a limit, so that any lowest amount of density is possible.
2.   The decrement follows a simple power law which follow the generic form $y = A\,x^\alpha$ with two parameters, A, the amplitude and α, the exponent

For our case, the amplitude A is chosen as unity, so one can write the power law simply as $\rho(L) = L^\alpha$. The slope of Fig. 8.5 gives the value of the exponent α, so

$$\text{Gradient} = \left[\log 1 - \log\left(\frac{3}{4}\right) \right] / (\log 1 - \log 2)$$

$$= \frac{\log 3}{\log 2} - 2 \tag{8.2}$$

Next, an important parameter, known as the fractal dimension (d_f) can be defined by the following equation:

$$M(L) = AL^{d_f} \tag{8.3}$$

With the substitution of (8.3) into (8.1), one has

$$\rho(L) = AL^d f^{-2} \tag{8.4}$$

Comparing (8.2) with (8.4), it shows that the Sierpinski Gasket has a fractal dimension $\frac{\log 3}{\log 2} = 1.58\ldots$, a dimension intermediate between that of a line (1D) and that of an area (2D).

8.6.2 Random Fractals

Unlike nonrandom fractals, random fractals [9] are found in nature. Examples are mountains, lighting, coastlines and clouds [10]. In the context of human anatomy, random fractals can be found in places like the regional distribution of pulmonary blood flow and mammographic parenchymal pattern as a risk for breast cancer. They are not necessarily geometrical in shape. In fact, they form a statistical average of some properties such as density which decreases linearly with length scale when plotted as a double logarithm paper.

The unbiased random walk problem of statistical mechanics will be used to illustrate the concept. At time $t = 0$, an ant is dropped onto an arbitrary vertex of an infinite one-dimensional lattice with constant unity or $x_{t=0} = 0$. An unbiased two-sided coin and a clock are carried by the ant. At each tick of the clock the ant flips the coin and heads towards the east, that is $x_{t=1} = +1$, if the coin indicates a head and vice versa or $x_{t=1} = -1$. As time processes, due to the law of nature, the average of the square of the displacement of the ant increases monotonically. The explicit form to express the increase concerning the mean square displacement can be shown by

$$\langle x^2 \rangle_t = t \tag{8.5}$$

Extending to higher powers of x, one can have

$$\langle x^k \rangle_t = 0 \tag{8.6}$$

for all odd integers of k and nonzero for all even integers of k. For example

$$\langle x^4 \rangle_t = 3t^2 - 2t = 3\,t^2\left[1 - \left(\frac{2}{3}\right)t\right] = \text{nonzero for } t = 0, 1, \ldots \qquad (8.7)$$

8.6.3 Other Definitions

From the comparison of Eqs. (8.5) with (8.6), one finds that the displacement of the random walking and can be defined as

$$L_2 = \sqrt{\langle x^2 \rangle} = \sqrt{t} \qquad (8.8)$$

Or

$$L_4 = \sqrt[4]{\langle x^4 \rangle} = \sqrt[4]{3} \cdot \sqrt{t}\left[1 - \left(\frac{2}{3}\right)t\right]^{1/4} \qquad (8.9)$$

This shows that the characteristic length L_2 and L_4 have the same asymptotic dependence on time. The leading exponent in the above equation is known as the scaling exponent and the nonleading exponent known as the corrections-to-scaling exponent. Extending to any length L_k, provided k is even, a general equation can be obtained as follows:

$$L_k = \sqrt[k]{\langle x^k \rangle} = A_k\sqrt{t}\left[1 + B_kt^{-1} + C_kt^{-2} + \cdots + O_kt^{-\frac{k}{2}+1}\right]^{1/k} \qquad (8.10)$$

The subscripts on the amplitudes indicate their k-dependence. The above equation displays the robust feature of random system. Also despite different definitions of characteristic length, the asymptotic behaviour is described by the same scaling exponent.

8.7 Fractal Approximations

Scientists at the Mount Sinai School of Medicine in New York City have successfully demonstrated that the fractal structure inside cells can indicate breast cancer [3]. Besides this, other works [6–8] have confirmed the chaotic nature of the scattering of ultrasound. Hence a fractal growth model can be used to adequately describe the scattered field within the breast.

From this, a new approximation method-the Fractal Approximation (FA) has been proposed by Leeman and Costa [11] based on the assumption that the scattered field

$u_s(\vec{r}')$ can be approximated by a modified version of the incident field $u_0(\vec{r}')$. The fractal growth probability distribution function $P(r, t)$ is introduced for the modification. The distribution function in turn is based on the fractal growth model known as the Diffusion Limited Aggregation (DLA) [12].

8.8 Diffusion Limited Aggregation

It has been discovered by scientists that many diverse natural phenomena have similar fractal shapes. That is, they are self-similar under different scaling factors. For instance, percolation clusters, a form of fractals are used to describe patterns created by water as it flows through coffee grinds or seeps into the soil. Comparable fractal patterns are also generated by the growth of some crystals and electrical charges.

Sander [12] and Thomas A. Witten in 1987 developed a model for fractal growth known as Diffusion Limited Aggregation (DLA). A random and irreversible growth process is used to create a particular type of fractal for their model. Today DLA has about 50 realizations in physical systems and much of current interest on fractals in nature are focused on FLA [12].

Aggregation is the process of growth of many clusters in nature. This means one particle after another comes into contact with a cluster and remains fixed in place. The resulting process is DLA if these particles diffuse toward the growing cluster along random walks [13].

Their research has the main assumption that the scattering paths of ultrasound in the breast follow the fractal-like structure of the DLAs. Therefore, one can model the internal scattered field $u_s(\vec{r}')$ with the fractal growth model.

To generate and grow a DLA cluster, first a seed particle is placed at the origin. This is followed by the release of random walkers one at a time, from some distant locations around the circumference of a circle surrounding the site of the origin.

When one of these particles makes contact with the seed at the origin, it sticks and forms an aggregate followed by the release of the next particle. The particle will be considered void and removed if it touches the boundary of the circle before reaching the origin. Self-similar clusters shown in Fig. 8.6 will be created by the totally random motion of the particles. It is noted that the dimension of the clusters increases as represented by s.

8.9 Growth Site Probability Distribution

From the description in previous section, it shows that DLA cluster is a form of probability. Here each step of a random walker is governed and described by a probability distribution known as the Growth Site Probability Distribution (GSPD) [15].

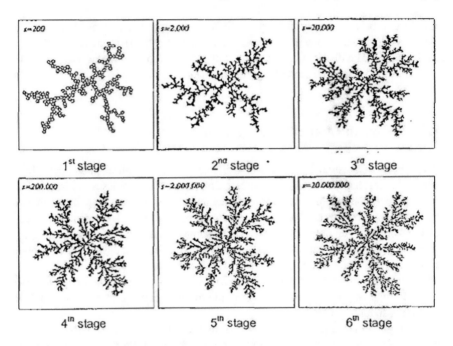

Fig. 8.6 A two dimension DLA clusters at six different stages of growth. After Sander [12]

There are a few types of GSPD in existence. In this chapter, one will consider the diffusion process of a random walker who is slowed down by entanglements such as dangling ends, large holes and bottlenecks [15]. This type is chosen because the scattering of ultrasound is also affected by the internal structure of the object which is inhomogeneous and the surrounding medium.

Averaging this distribution over all the starting points of the random walkers, which are the coordinates of each sample point in the projection of the forward-scattered field, the probability distribution reduces to a stretched version of the Gaussian distribution [15]. Then the distribution $\langle P(r, t) \rangle$ is given by the following expression:

$$\ln\left(\frac{\langle P(r, t) \rangle}{\langle P(0, t) \rangle} \right) \sim -\left(\frac{r}{\langle r^2(t) \rangle^{1/2}} \right)^{\frac{d_w}{d_w - 1}} \tag{8.11}$$

where $\langle P(0, t) \rangle$ = the average probability of finding a random walker at the starting point, $\langle r^2(t) \rangle^{1/2}$ = the root mean square distance of the walker from the starting point, and d_w = fractal dimension of a random walk or the diffusion exponent.

There is a strong evidence that Eq. (8.11) is valid for a large class of random fractals [6–8]. From the above equation, the average GSPD, $\langle P(r, t) \rangle$ is given by:

$$\langle P(r,t) \rangle \sim \langle P(0,t) \rangle \exp\left[-\left(\frac{r}{\langle r^2(t) \rangle^{1/2}}\right)^{\eta}\right] \qquad (8.12)$$

where $\eta = \frac{d_w}{d_w - 1}$.

With a long time span, the average probability at the starting point, $\langle P(0,t) \rangle$ is proportional to the inverse of the number of distinctly visited sites, $S(t)$. $S(t)$ scales as $\langle r^2(t) \rangle^{d_f/2}$ on fractals, and one has

$$\langle P(0,t) \rangle \sim \left(\frac{1}{\langle r^2(t) \rangle}\right)^{d_f/2} \sim t^{-d_f/d_w} \qquad (8.13)$$

where d_f = fractal dimension of the DLA cluster, and d_w = the diffusion exponent.

It is mentioned earlier that the motion of the random walker is assumed to be slowed down by entanglements. Hence the root mean square distance of the walker is given by a more general power law [6–8] as:

$$\langle r^2(t) \rangle \sim t^{2/d_w} \qquad (8.14)$$

With the substitution of (8.14), (8.13) into (8.12), the GSPD is then given by

$$\langle P(r,t) \rangle = t^{-d_w/d_f} \cdot \exp\left[-\left(\frac{r}{\left(t^{\frac{2}{d_w}}\right)^{\frac{1}{2}}}\right)^{u}\right] \qquad (8.15)$$

This shows that there are two unknown parameters which have to be defined: d_w, the diffusion coefficient and d_f, the fractal dimension of the DLA cluster.

8.10 Approximating of the Scattered Field Using GSPD

With the general equation of GSPD, it can be applied to the modelling of the ultrasound scattered field within the breast. In diffraction tomography, the general expression for the incident field, assuming plane wave, is given by

$$u_0(\vec{r}) = A e^{j\vec{k}\cdot\vec{r}} = A e^{j(k_x x + k_y y)} \qquad (8.16)$$

where A = amplitude of the incident ultrasound field, $\vec{k} = (k_x, k_y)^T$ with $k_x^2 + k_y^2 = k^2$, $\vec{r} = (x, y)^T$ and k = wave number.

Multiplying the incident wave field with the GSPD, the scattered ultrasound field within the object is

$$u_{SF}\left(\overrightarrow{ri}\right) = Ae^{j\left(k_x x' + k_y y'\right)} \cdot t^{-d_w/d_f} \cdot \exp\left[-\left(\frac{R}{\left(t^{\frac{2}{d_w}}\right)^{\frac{1}{2}}}\right)^{d_w/(d_w-1)}\right] \tag{8.17}$$

where R = absolute distance between the coordinates of a scatterer within the breast $(\overrightarrow{ri} = (x', y')^T$ and a sampling point, the starting point of random walk.

In this work, A is taken to be unity, and the value of t for every scatterer is taken to be 1. The average GSPD $\langle P(R, t)\rangle$ is obtained by taking the average of each probability value obtained for all R, that is, all scatterers with respect to a sampling point. The total ultrasound field within the object is given by the summation of the incident field and the scattered field:

$$u\overrightarrow{(r')} = u_0\overrightarrow{(r')} + u_{sF}\overrightarrow{(r')}$$

$$= Ae^{j\left(k_x x' + k + y'\right)} + Ae^{j\left(k_x x' + k_y y'\right)} \cdot t^{-d_w/d_f} \exp\left[-\left(\frac{R}{\left(t^{\frac{2}{d_w}}\right)^{\frac{1}{2}}}\right)^{d_w/(d_w-1)}\right]$$

$$\tag{8.18}$$

By substituting the total field into the Lippmann–Schwinger integral [16], the final expression is given by

$$u_s(\vec{r}) = \int_g \left(\vec{r} - \overrightarrow{r'}\right)n\left(\overrightarrow{r'}\right)\left[u_o\left(\overrightarrow{r'}\right) + u_{SF}\left(\overrightarrow{r'}\right)\right]\mathrm{d}\overrightarrow{r'} \tag{8.19}$$

where $n\overrightarrow{(r')}$ = object function, $u_s\overrightarrow{(r')}$ = scattered field.

So there is no need to omit the scattered field $u_s\overrightarrow{(r')}$ to solve the Lippmann–Schwinger equation. This scattered field plays a significant role in the determination of the image resolution.

In diffraction tomography, the next step is to derive the reconstruction algorithm to solve for the object function and generate the image. This is a tedious process as the higher the image resolution, the more object functions will be needed. Hence one has to develop an efficient reconstruction algorithm to solve the equations represented in matrix form.

8.11 Discrete Helmholtz Wave Equation

Here the Helmholtz wave equation is solved by first converting it in discrete form. This discrete equation then expresses each sample of the projection data as a summation of all scatterers within the object:

$$u_s(\vec{r}) = \sum_r g\left(\vec{r} - \vec{r'}\right)u(\vec{r'})n(\vec{r'}) \qquad (8.20)$$

After taking all the samples of each projection into considerations, a vector equation will be derived from the above equation as

$$U = AN \qquad (8.21)$$

where

.

$$\begin{bmatrix} U = [u_s(r_1)u_s(r_2)\ldots u_s(r_n)^T \\ N = [n(r'_1)n(r'_2)\ldots n(r'_m)]^T \\ A = g(r_1 - r'_1)u(r'_1) \quad g(r_1 - r'_m)u(r'_m) \\ g(r_n - r'_1)u(u'_1) \quad g(r_n - r'_m)u(r'm) \end{bmatrix}$$

where $n = (1, 2,...) =$ total number of projections samples and $m = (1, 2, ...)$ is total number of scatterers.

The vector U, the projection scattered field can be simulated or measured. For this chapter, it will be simulated. The simulation details will be given in the next section.

The matrix A contains products of the total field and Green's function. The generated image will represent the object in its surrounding medium. Hence the values of $\vec{r'}$ are coordinates of both points located in the surrounding medium and the scatterers. A circle of radius R is the boundary of the object. This circle is slightly bigger than the object to be examined. In this way all the scatterers found within this circle are considered to be the object itself. The reason of using a circle is that the interior view of the breast is circular in shape. The total field within this circle will be given by the summation of the scattered field, calculated through fractal approximation. Also the field outside the circle will consist only of the incident field.

The vector N has to be solved. It contains unknown values of both the object function and the surrounding medium. Equation (8.21), a linear algebraic equation can be solved conventionally using the LU decomposition [17] or the Gaussian elimination method when the number of projections and samples involved is small, for instance 32 projections by 32 samples or less. However, such methods are not applicable. When the image resolution is 64×64 or above due to the size of matrix A with 4096 rows by 4096 columns is too large for storage in programs like Matlab.

8.12 Kaczmarz Algorithm

When solving equation with large matrices, usually it is assumed that a significant number of elements within the matrix are zero. So that sparse matrix techniques can be used.

One possible method of solving large matrices with nonzero elements is the use of the Kaczmarz algorithm [18]. In this algorithm, the problem of memory wastage is solved by operating only one row of the matrix at a time.

This algorithm also guarantees a proper solution of the linear algebraic equation (8.21) with proper convergence. Hence it is able to satisfy the Helmholtz wave equation with proper discretization of all the functions. The Kaczmarz algorithm uses the technique of solving the linear equation $U = A.N$ by representing each row of the vector equation by a separate equation.

For a set of equation consisting of one projection by three samples, it is written as

$$u_s(\vec{r_1}) = g\left(\vec{r_1} - \vec{r_1'}\right)u\left(\vec{r_1'}\right)n\left(\vec{r_1'}\right) + g\left(\vec{r_1} - \vec{r_2'}\right)u\left(\vec{r_2'}\right)n\left(\vec{r_2'}\right)$$
$$+ g(\vec{r_1} - \vec{r_3'})u\left(\vec{r_3'}\right)n\left(\vec{r_3'}\right)$$
$$u_s(\vec{r_2}) = g\left(\vec{r_2} - \vec{r_1'}\right)u\left(\vec{r_1'}\right)n\left(\vec{r_1'}\right) + g\left(\vec{r_2} - \vec{r_2'}\right)u(\vec{r_2'})n\left(\vec{r_2'}\right)$$
$$+ g(\vec{r_2} - \vec{r_3'})u\left(\vec{r_3'}\right)n\left(\vec{r_3'}\right)$$
$$u_s(\vec{r_3}) = g\left(\vec{r_3} - \vec{r_1'}\right)u\left(\vec{r_1'}\right)n\left(\vec{r_1'}\right) + g\left(\vec{r_3} - \vec{r_2'}\right)u\left(\vec{r_2'}\right)n\left(\vec{r_2'}\right)$$
$$+ g(\vec{r_3} - \vec{r_3'})u\left(\vec{r_3'}\right)n\left(\vec{r_3'}\right) \tag{8.22}$$

Each of the above equations represents a hyperplane in three-dimensional space. The solution is given by the intersection of these planes. The method of the Kaczmarz algorithm by projecting its point onto another hyperplane in sequence to refine an initial guess. Each step of the iteration enables the new point to get closer to the final solution. With each row of matrix A represented by a_i where i = row number, each projection sample represented by u_i and each row of the object function represented by n_k, where k = the iteration number, a better estimate of $u_i = a_i.n_k$ is given by

$$n_{k+1} = n_k - [(a_i, n_k - u_i)/(a_i \cdot a_i)]a_i \tag{8.23}$$

The following example illustrates the convergence of the Kaczmarz algorithm as shown in Fig8.7. This shows hyperplanes in two-dimensional space. Here the initial guess n, on line AB is first projected onto hyperplanes CD giving $a_{11}n_1 + a_{12}n_2 = u_1$. A new Estimate n_2 is produced on line CD. With the projection of this new estimate onto the line EF, the next new estimate n_3 is produced.

One notes that each estimate is getting closer and closer to the solution of these equations which is the intersection point of these two hyperplanes.

Fig. 8.7 Convergence of estimates in Kaczmarz [18] Algorithm. After Kaczmarz [18]

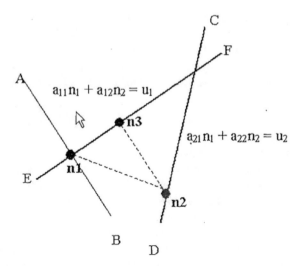

8.13 Hounsfield Method

The Kaczmarz algorithm guarantees convergence. However, it has the major concern on the speed of convergence. The interdependency of the row equations is a factor that determines the speed of convergence. If the hyperplanes are perpendicular to each other, the correct answer can be calculated by only one iteration. On the other hand, if they are parallel to each other, this produces very slow speed of convergence and more iterations will be needed. In this case, one needs to orthogonalize the equations to speed up the convergence. However, this method requires a similar amount of storage space and work to find the inverse of matrix A.

Ramarkrishnan [19] proposed a less computationally expensive method, the pairwise orthogonalization method. This method orthogonalized a hyperplane to its previous hyperplane by the following relations:

$$\widetilde{A}_i = A_i - \widetilde{A}_{i-1}\left[\left(\widetilde{A}_i.A_{i-1}\right)/\left(\widetilde{A}_{i-1}\widetilde{A}_{i-1}\right)\right] \tag{8.24}$$

and

$$\widetilde{b}_t = b_i - \widetilde{b}_{i-1}\left[\left(\widetilde{A}_t \cdot A_{i-1}\right)/\widetilde{A}_{i-1} \cdot \widetilde{A}_{i-1}\right] \tag{8.25}$$

This new orthogonal system of equations is represented by $\widetilde{A}_i$ and $\widetilde{b}_i$ respectively. This method reduces the need for large storage space as only an additional equation is needed at one time although it is not optimum. In fact it has been proven that it would actually reduce the number of iterations by half.

Hounsfield [20] introduced another alternative method by rearranging the order of equations to reduce the interdependency between adjacent equations. It stated that convergence is slow due to the fact that the hyperplane at adjacent points are usually parallel to each other. Convergence would be faster if parallel hyperplanes are saved until later for further iterations. This can be done by rearranging the equations.

The degree of interdependence of two equations can be determined by the angle between two hyperplanes. The angle is nearly zero for parallel equations. The angle is 90° for perpendicular equations. The angle between two hyperplanes is definitely

$$\cos\theta = A_i.A_j/\sqrt{(A_i.A_i)(A_j.A_j)} \qquad (8.26)$$

where A_i, A_j = rows in matrix A.

It was found that the convergence is usually larger if the average angle values between equations are larger. This is usually true for objects with smaller refractive indices. One can also increase the angle between hyperplanes by skipping a number of equations between each calculation and the speed of convergence can be increased [20]. Comparing the Hounsfield method with that of Ramakrishnan [19], it is equally effective in speeding up the convergence but not the work of calculation. In this chapter, this method will be use with some modification in order to increase the rate of convergence.

Besides this, the rate of convergence will also be affected by the sampling interval. It is found that with smaller sampling interval for objects with large refractive indices, the rate of convergence will improve.

8.14   Applying GSPD into Kaczmarz Algorithm

In previous section, it is known that by reducing their interdependency through the rearrangement of equations, the rate of convergence will increase. Ideally by grouping together those equations with hyperplanes that are separated by the largest angles an optimum arrangement can be achieved. However, it is of tremendous computational work involved with to find a pair of equations that is least interdependent. This is because each hyperplane will have to be measured against all other hyperplanes.

This can be done by exploiting the useful relationship between the sampling points and the scatterers as expressed in the GSPD.

A suboptimum arrangement with minimum work can be obtained by rearranging the equations according to the steps shown below.

First by focusing on a scatterer at a time, the first two rows of an imaginary matrix can be formed by grouping together two rows of equations containing the scatterer's highest and lowest probabilities in relation to a sampling point. The chance of having two adjacent hyperplanes is minimum by rearranging the equations in this way. This can improve the rate of convergence because the angle between these two hyperplanes is also large. The equation can also be suboptimized by permutating the sequence of equation randomly if the above procedure is not possible due to the lack of relationships.

The above equation will generate the object function. Once the object function is known, its amplitude values, real or imaginary are then multiplied by the colour or gray scale factor to produce the pixel values of the image. Then a tomographic image can be obtained based on these pixel values

8.15 Fractal Algorithm using Frequency Domain Interpretation

The method in the previous section using the space domain interpolation for the successful reconstruction of a tomographic image is too time-consuming to be used in practical medical ultrasound imaging system. Hence, we here propose an alternative frequency domain interpolation to be implemented with the fractal approximation of the scattered field. The following section provides the background of the algorithm used.

8.16 Derivation of Fractal Algorithm's Final Equation Using Frequency Domain Interpolation

The conventional diffraction tomography treatment uses the Born approximation in the Lippmann-Schwinger integral equation.

This is a first order approximation, treating the object as a point object and neglecting the scattering within the object. Here instead of substituting Born's approximation into the Lippmann-Schwinger integral equation, the total field is substituted instead.

Equation (8.18), the total field is given by

$$\vec{u}(\vec{r}') = Ae^{j(k_x x + k_y y)}\left[1 + t^{-d_w/d_f}\right] \cdot \exp\left[-\left(\frac{R}{\left(t^{\frac{2}{d_w}}\right)^{\frac{1}{2}}}\right)^{d_w/(d_w-1)}\right]$$

Since d_f, d_w and the distance R are constants at any specific time t,

$$1 + t^{-d_w/d_f} \cdot \exp\left[-\left(\frac{R}{\left(t^{\frac{2}{d_w}}\right)^{\frac{1}{2}}}\right)^{d_w/d_w-1}\right]$$

is also a constant at any specific time t.

$$\text{Let } \mu = 1 + t^{-d_w/d_f} \cdot \exp\left[-\left(\frac{R}{\left(t^{\frac{2}{d_w}}\right)^{\frac{1}{2}}}\right)^{d\underline{w}/(d_w-1)}\right] \tag{8.27}$$

substituting (8.27) into equation (8.18), the total field,

$$\vec{u}(\vec{r}') == u_0(\vec{r}') + u_{SF}(\vec{r}') = \mu \cdot A \cdot \exp(k_x x + k_y y) \tag{8.28}$$

Assuming $A = 1$, in the frequency domain,

$$u_B(\alpha, L_0) = \mu^* \frac{j}{2\beta} e^{j\beta L_0} O(\alpha, \beta, -k) \tag{8.29}$$

By using conventional diffraction tomography's frequency domain interpolation procedure on Eq. (8.29), the object function, $n(\vec{r})$ can be obtained in a much shorter time.

8.17   Simulation Results

The diffraction tomography experiment is to be carried out at a frequency of 5 MHz with an ultrasound wavelength in water of 0.3 mm [20]. The parameters used in the simulations are based on a real situation. The simulated images are given in Figs. 8.8, 8.9 and 8.10.

Figure 8.9 shows an improvement in the image resolution compared to that using Born approximation.

$$\text{Phantom} = \begin{bmatrix} A_{1\times1} & A_{1\times64} \\ A_{64\times1} & A_{64\times64} \end{bmatrix}$$

$$\text{Born Approximation} = \begin{bmatrix} B_{1\times1} & B_{1\times64} \\ B_{64\times1} & B_{64\times64} \end{bmatrix}$$

$$\text{Fractal Approximation} = \begin{bmatrix} C_{1\times1} & C_{1\times64} \\ C_{64\times1} & C_{64\times64} \end{bmatrix}$$

$$\text{RMS Born Approximation} = \sqrt{\sum_{j=64}^{i=64} [(B_{ij} - A_{ij})^2]/i * j} = 0.20812 \tag{8.30}$$

$$\text{RMS Fractal Approximation} = \sqrt{\sum_{j=64}^{i=64} \frac{(C_{ij} - A_{ij})^2}{i * j}} = 0.0034232 \tag{8.31}$$

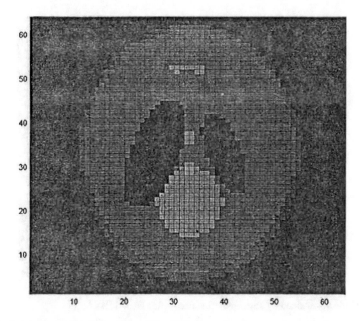

Fig. 8.8 Original phantom. After Jia and Han [14]

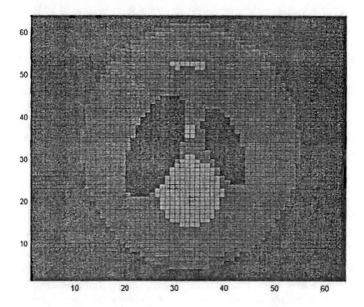

Fig. 8.9 Image obtained by fractal approximation. After Jia and Han [14]

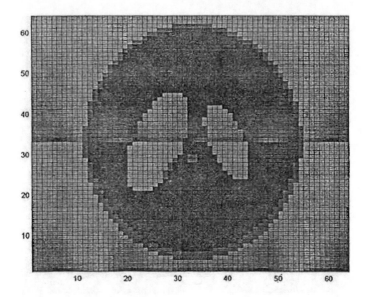

Fig. 8.10 Image obtained by born approximation. After Jia and Han [14]

The root mean square values of the pixels of the images formed by Born approximation and fractal approximation are given by Eqs. (8.30), (8.31). The fractal approximation shows a smaller value with respect to the original image. This shows that the fractal approximation is a better technique. Generally, Figs. 8.11, and 8.12 show images obtained by fractal approximation of Hankel function.

K0 is chosen as 1 or 2 in the Bessel function of Third Order-Hankel function [21]. If K0 is changed to 300, Fig. 8.11 shows that the image obtained is distorted. The main ellipses within the phantom have been ignored during the calculation of the Bessel function. A value of 1 or 2 is suggested for K0 because Fig. 8.12 is the mesh plot of Fig. 8.11.

8.18   Comparison Between Born and Fractal Approximation

In general, images from Born approximation and fractal approximation both suffer from severe artefacts. There are reasons for this. One is the number of iterations is insignificant in bringing the initial guess of the object function closer to the final result. The second reason is due to the image resolution is based only on 64 × 64 projection samples. This is considered very low in resolution for medical or image processing standard. Thus artefacts can be reduced with more rounds of iterations and higher resolution [3]. The differences between Born approximation and fractal approximation are still noticeable in the images although the limited rounds are

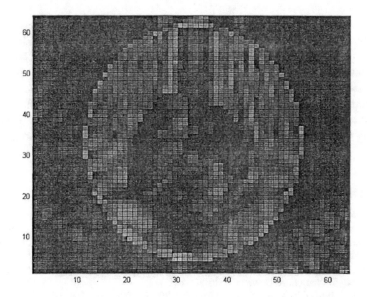

Fig. 8.11 Image obtained by fractal approximation of Hankel function K0 = 300, P colour Plot. From Jia and Han [14]

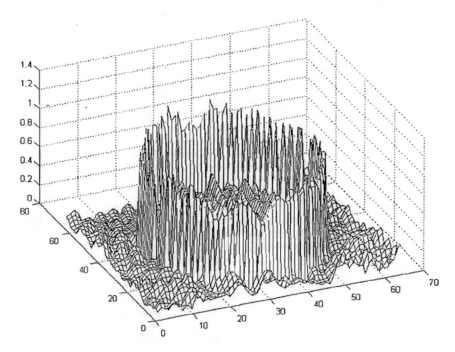

Fig. 8.12 Image obtained by fractal approximation of Hankel Function K0 = 300, Mesh Plot. From Jia and Han [14]

iterations are not able to bring out the progressive convergence of the Kaczmarz algorithm. An obvious difference is shown in the presence of the boundary of the main ellipse in the images based on the fractal approximation. Also the boundary of the object is established at such an early stage of iteration of the fractal approximation. This marks a clear improvement over the Born approximation. The reduction of the magnitude of the artefacts is within the object is another improvement. This is shown by the intensity of the colour within the object's boundary reduces whereas the colour intensity of images remains unchanged when based on the Born approximation [3].

References

1. Gan, Woon Siong. 1992. Application of chaos to sound propagation in random media. In *Acoustical Imaging*, vol. 19,99–102,ed. H. Ermert, and H. Harjes. New York: Plenum Press.
2. http://library.thinkquest.org/3120/test/c-his1.htm.
3. Chow, W.H. 2002. *Acoustic Fractal Imaging Using Diffraction Tomography*. M.Eng. thesis, Nanyang Technological University, chapter 5, p. 112.
4. Gleick, J. 1981. *Chaos-Making A New Science*, 29. New York: McGraw Hill.
5. Laplante, P., Fractal Mania, Windcrest/McGraw Hill, USA.
6. Gan, W.S., and C.K. Gan. 1993. *Acoustical Fractal Images Applied to Medical Imaging, Acoustical Imaging*, vol.20, 413–416. Plenum Press, New York.
7. Lam, C.K., and E.H. Lau. 1996. *Acoustical Chaotic Image for Medical Imaging*. B.Eng. Thesis, NTU.
8. Kjems, J.K. 1996. Fractals and experiments. In *Fractals and Disordered System*, 2nd ed., 284–286. Berlin: Springer.
9. Stanley, H.E. 1996. Fractals and multifractals: the interplay of physics and chemistry. In *Fractals and Disordered Systems*, 2nd ed., 1–13. Berlin: Springer.
10. http://members.home.net/jason.jiin/content/.htm.
11. Leeman, S., E.T. Costa. 1993. Large aperture hydrophones for far field measurement and calibration, IEE, London. *Acoustic Sensing & Imaging*, 294.
12. Sander, L.M. 1987. *Fractal Growth*, 94. Scientific American.
13. Vicsek, T. 1989. *Fractal Growth Phenomena*,1st ed, 158–167. Singapore: World Scientific.
14. Jia, Yu, and Chan Tuck Han. 2003. *Development of an Ultrasound Imaging Algorithm for Detecting Breast Cancer*. B.Eng. Thesis, NTU.
15. Bunde, A., and S. Havlin (eds.). 1996. *Fractals and Distorted Systems*, 2nd ed., 7. Berlin: Springer.
16. Kaveh, M., M. Soumekh, and R.K. Mueller. 1986. A comparison of Born and Rylov approximatin in acoustic tomography. *Acoustical Imaging* 14: 325–334.
17. http://www.nr.com/Numerical Recipes in Chapter 2.
18. Kaczmarz, S. 1937. Angenäherte Auflösung von Systemen Linearer Gleichungen. *Bulletin Academie Polonaise Sciences et Lettres A* 355–357.
19. Ramakrishnan, R.S., S.K.J. Mullick, R.K.S. Rathore, and R. Subramanian. 1979. Orthogonalisation, Bernstein polynomials and image restoration. *Applied Optics* 18: 464–468.
20. Hounsfield, G.N. A method of apparatus for examination of a body by radiation such as x ray or gamma radiation. Patent Specification, The Patent Office.
21. Slaney, M., and A.C. Kak. 1985. *Imaging with Diffraction Tomography*, 39. University of Purdue Internal Report, TR-EE 85-5.

Chapter 9
Nonclassical Nonlinear Acoustical Imaging

9.1 Introduction

A traditional view of classical nonlinearity of nonlinear ultrasonics is based on the idea of elastic waveform distortion due to material nonlinearity. This waveform distortion is due to the variation in local sound velocity as the sound wave propagates in the material. This sound velocity variation is accumulated with propagation distance resulting in the progressive transition of the incident harmonic wave into the sawtooth or N-type wave. The consequence is that the spectrum acquires higher harmonics of the fundamental frequency which provide information on the material. For sound propagation in free-from-defects media, the material nonlinearity is rather low and a few harmonics are observable. Hence classical nonlinear nondestructive testing (NDT) is basically second harmonic NDT.

Nonclassical nonlinear acoustical imaging involves nonclassical contact acoustic nonlinearity (CAN) [1] spectra. Nonclassical nonlinear acoustical imaging has applications in material characterization and nondestructive testing. In these two applications, two forms of nonclassical acoustic nonlinearity are involved with, the mesoscopic or hysteretic nonlinearity and contact acoustic nonlinearity (CAN). Mesoscopic nonlinearity originates from the bond system of inclusions like dislocations, cracks, grain contacts etc. and in heterogeneous materials. This results in a hysteretic stress–strain and strongly nonlinear relation. The nondestructive testing applications are related to the study of the slow dynamic behaviour of the hysteretic materials. This is a strong mechanical impact followed by a slow recovery. The latter is an indication of the presence of defects. It is robust methodology for a pass/nonpass evaluation in nondestructive testing.

The contact acoustic nonlinearity (CAN) is caused by the mechanical constraint between the fragments of planar defects which make their vibrations extremely nonlinear. CAN in general manifests in a wide class of damaged materials. Multiple ultraharmonics are efficiently generated by the cracked defects which also support multiwave interactions. The planar defects also have resonance properties bringing in nonlinear resonance with ultra-subharmonics (USB) spectra into elastic wave-defect interactions. This is another contribution to the nonlinear vibration spectrum.

9.2 Mechanisms of Harmonic Generation Via Contact Acoustic Nonlinearity (CAN)

9.2.1 Clapping Mechanism

First a prestressed crack with static stress σ^0 is considered. It is driven with longitudinal acoustic traction σ_2. This traction is sufficiently strong to cause clapping of the crack interface. The asymmetrical dynamics of the contact stiffness produces the clapping nonlinearlity. This asymmetrical dynamics, due to clapping is higher in a compression phase than that for tensile stress when the crack is assumed to be supported only by an edge-stress.

The following stress (σ)-strain (ε) relation [2] can be used to describe approximately the above behaviour of a clapping interface:

$$\Sigma = C\left[1 - H(\varepsilon)\left(\frac{\Delta C}{C}\right)\right]\varepsilon \tag{9.1}$$

where C = contact material stiffness, $\Delta C = [C - \left(\frac{d\sigma}{d\varepsilon}\right)_{\varepsilon > 0}]$, and $H(\varepsilon)$ = Heaviside unit step function.

The above bimodular pretressed contact driven by the harmonic strain $\varepsilon(t) = \varepsilon_0 \cos\nu t$ is similar to a mechanical diode. This results in a pulse-type modulation of its stiffness $C(t)$ as shown in Fig. 9.1.

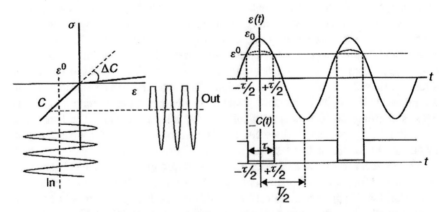

Fig. 9.1 Mechanical diode model (left); stiffness modulation and waveform distortion (right). After Solodov [9]

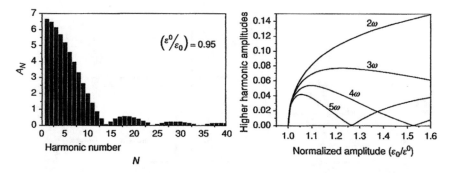

Fig. 9.2 CAN higher harmonic spectrum (left) and dynamic characteristics (right). After Solodov [9]

An unconventional nonlinear waveform distortion is produced. It is a half-period rectified output instead of the sawtooth-like waveform produced by classical materials' nonlinearity (Fig. 9.1).

Due to the fact that $C(t)$ is a pulse-type periodic function of the driving frequency ν (Fig. 9.1, right), the nonlinear part of the spectrum induced in the damaged area of $\sigma^{NL}(t) = \Delta C(t)$. $\varepsilon(t)$ shall contain a number of its higher harmonics $n\nu$ with both even and odd orders. Their amplitudes are modulated by the sinc-envelope function [2]:

$$A_n = \Delta C \Delta \tau \varepsilon_0 [\sin C(n+1)(\Delta \tau) - 2 \cos(\pi \Delta \tau) \sin C(n \Delta \tau) + \sin C((n-1)\Delta \tau)]$$
$$(9.2)$$

where $\Delta \tau = \tau/T(\tau = T/\pi)\text{arc}\cos(\varepsilon^0/\varepsilon_0)) =$ normalized modulation pulse length.

Equation (9.2) produces the spectrum of the nonlinear vibration and is shown in Fig. 9.2 left.

This spectrum contains a number both even and odd higher harmonics arising simultaneously as soon as $\varepsilon > \varepsilon^0$ (the threshold of clapping). In Eq. (9.2), the sinc-modulation is amplitude dependent. This is shown as the wave amplitude ε_0 increases, T grows from o to $T/2$. The dynamic characterization of the higher harmonics will be affected by this as shown in Fig. 9.2 right. Due to the spectrum compression effect, this also provides their oscillation.

9.2.2 Nonlinear Friction Mechanism

The surfaces of the contacts interface are mechanically coupled by the friction force caused by the interaction between asperities due to the shear wave drive. The interface shear motion is constrained by the interaction between neighbouring asperities if the driving amplitude is small enough. This prevents the contact surfaces from sliding or the micro-slip mode.

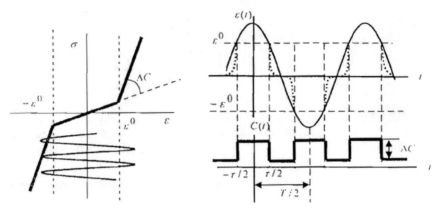

Fig. 9.3 Mechanical diode model (left), stiffness modulation and waveform distortion in micro-slip mode (right). After Solodov [9]

Figure 9.3 shows the mechanical diode model [3] for the micro-slip motion. It demonstrates a stepwise increase in tangential stiffness as the neighbouring asperities interact. This causes a stiffness variation twice for the input signal period due to this interaction which is independent of the shear motion. This is shown in Fig. 9.3, right. This provides only odd harmonic generation and such a constraint also introduces a symmetrical nonlinearity. Their amplitudes are sinc-modulated which is similar to the clapping mechanism:

$$A_{2N+1} = 2\,\Delta C \varepsilon_0 \left(\frac{\tau}{T}\right) \left\{ \sin C \left(\frac{2N\tau}{T}\right) + \sin C \left(\frac{2(N+1)\tau}{T}\right) \right\} \qquad (9.3)$$

This is shown in Fig. 9.4.

It also exhibits similar nonpower dynamics. Static friction for the motion-slip motion changes for the sliding when the amplitude of tangential traction is greater than the contact. The contact stress–strain relation follows a hysteresis loop [1] due to an oscillating shear wave drive is accompanied by a cyclic transition between static and kinematic friction or the stick-and-slide mode. The contact tangential stiffness changes symmetrically between the static for a stick phase and dynamic for a slide phase values twice over the input strain period, independent of the direction of motion. This provides the generation of odd higher harmonics similar to the contact acoustic nonlinearity (CAN) features of nonpower dynamics and sinc-spectgrum modulation.

9.3 Nonlinear Resonance Modes

The experiment of Solodov et al. [4, 5] discovered that besides generating higher harmonics, there are other scenarios of contact acoustic nonlinearity (CAN) which

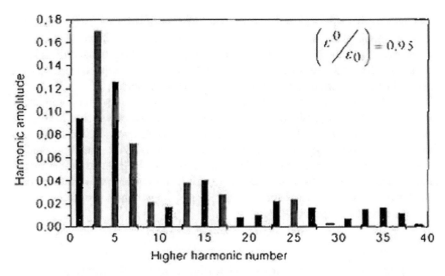

Fig. 9.4 Sinc-modulated odd higher harmonic CAN spectrum in micro-slip mode. After Solodov [9]

considerably expands the nonlinear spectrum of cracked defects. This additional scenarios is chaos phenomenon which is an abrupt change of output for a slight variation in the input parameter. This is a form of dynamic instability.

Solodov et al. [4] illustrated the feasibility of the new vibration modes and ascertain their basic spectra pattern, by assuming that the crack exhibits both nonlinear properties and resonance. They identified the crack as a nonlinear oscillator. The characteristic frequency (ω_0) of the oscillator is given by an associated mass of material inside the damaged area and a linear stiffness. A displacement-dependent (X) nonlinear interaction force $F^{NL}(X)$ represents the contact nonlinearity. The driven vibration with a driving force $f(t) = f_0 \cos \nu t$ of the nonlinear oscillator is given by the solution of the following nonlinear equation:

$$\ddot{X} + \omega_0^2 X = f(t) + F^{NL}(X) \qquad (9.4)$$

where X = displacement.

A second-order perturbation approach is used. That is $F^{NL} \sim \cos(\nu - \omega_0)$. This accounts for the interaction between the driving and natural frequency vibrations. A resonance increase in the output at $\omega_0 \approx \nu/2$ is observed if $\nu - \omega_0 \approx \omega_0$. This is subharmonic generation. The frequency relation $m\nu - n\omega_0$ provides the higher-order terms in the interaction. This gives rise to a resonance output at $\omega_0 \approx m\nu/(n + 1)$. The crack generation an ultra-subharmonic (USB) of the second order $m\nu/2$ for $n = 1$. Higher values of n produce higher-order ultra-subharmonics (USB). A damaged area has a more complicated structure than a set of coupled nonlinear oscillators for real situation. A cross-excitation of the coupled oscillation can be provided by a frequency of the driving acoustic wave as $\nu = \omega_\alpha + \omega_\beta$ with the difference frequency components of $v - \omega_\alpha = \omega_\beta$ and $v - \omega_\beta \approx \omega_\alpha$. The result is a resonant generation of the frequency pair, ω_α, ω_β, centred around the subharmonic position. According to Solodov et al. [4], the higher-order nonlinear terms in Eq. (9.4) expand the contact acoustic nonlinearity (CAN) spectrum, which comprises a multiple ultra-frequency pairs (UFP) centred around the USB and the higher harmonics.

The USB and UFP can be interpreted as the half-frequency and combination frequency decay respectively of a high-frequency (HF) phonon which is the driving frequency signal. USB and UFP belong to a class of instability mode. The avalanche-like amplitude growth beyond the input threshold is due to this resonance instability. The input amplitudes for the up and down transitions are different; known as amplitude hysteresis. This is a form of bistability [5] due to the reverse amplitude excursion. This form of dynamics is a distinctive signature of the nonlinear resonance, a form of nonlinear acoustic phenomena.

9.4   Experimental Studies on Nonclassical CAN Spectra

Solodov et al. [6] used nonlinear laser vibrometry (NLV) [6] to study the nonclassical CAN spectra. Their results are given in Figs. 9.5 and 9.6.

The USB spectrum in the cracked area of a polystyrene plate with a driving frequency of around 13 kHz with a shaker is shown in Fig. 9.5. The involvement of the CAN mechanism into USB generation is shown by the way amplitude modulation. The UFP spectrum in an impact-damaged area of GFRP spectrum with a driving frequency of 20 kHz is shown in Fig. 9.6. The position of the second (40 kHz) and third (60 kHz) harmonics as well as USBs (50 and 70 kHz) can be identified.

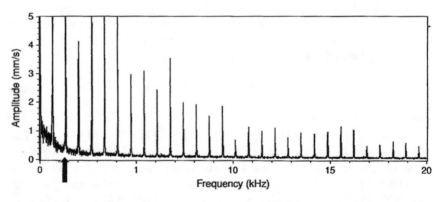

Fig. 9.5 USB spectrum in cracked area of polystyrene plate. The arrow indicates driving frequency. After Solodov [9]

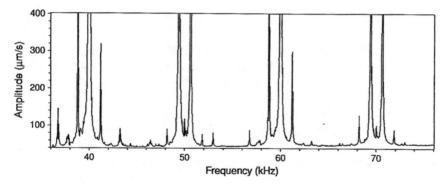

Fig. 9.6 UFP spectrum in impact damaged area of GFRP-specimen. Driving frequency Is 20 kHz. After Solodov [9]

9.4.1 CAN Application for Nonlinear Acoustical Imaging and NDE

The intact part of material outside the defects vibrate linearly, with no frequency variation in the output spectrum, while the nonlinear spectra in Figs. 9.5 and 9.6 are produced locally in the damaged area. Hence nonlinear defects are active sources of new frequency components and not passive scatterers in conventional ultrasound testing. Nonlinearity is thus a defect-selective indicator of the presence of damage. The nonlinear imaging of damage has its basis in the high localization of nonlinear spectral components around the origin. A sensitive scanning Laser interferometer is sued in NLV for the detection of the nonlinear vibration of defects.

Piezo-stack transducer operating at 20 and 40 kHz are used in the excitation system. The C-scan images of the sample area can be obtained for any spectral line within the frequency bandwidth of 1 MHz after a 2D scan and FFT of the signal received.

The image of an oval examination on top of a piezoactuator embedded into a GFRP composite is shown in Fig. 9.7. It shows smart structures which can be used for active

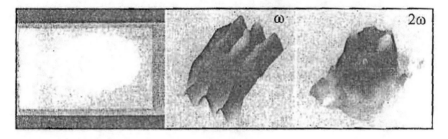

Fig. 9.7 Fundamental frequency (ω) and higher harmonic imaging of a delamination in a "smart" structure. After Solodov [9]

structural health monitoring for aerospace components. Such smart structures can be used.

The internal excitation source is the actuator itself. It is fed with an input of only a few volts. The boundary ring of the delamination is selectively shown by the higher harmonic images. This is where rubbing and clapping of the contact surface are expected. On the contrary, only a standing wave pattern over the area of the actuator is indicated by the driving frequency (50 kHz) image.

A fatigue cracking produced by cyclic loading in an Ni-based super-alloy is clearly shown by the USB image in Fig. 9.8a. The crack has a length of only around 1.5 mm and the average distance between the edge is only around 5 μm.

Linear NDE with a slanted ultrasonic reflection is unable to detect such small cracks. The UFP components, generally display strong spatial localization around the defects. These are applicable for the detection of damage. This is similar to all nonlinear modes.

Figure 9.9 illustrates the benefit of UFP mode for a 14-ply epoxy-based GFRP composite with a 9.5 J-impact damage in the central part.

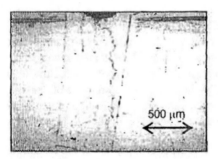
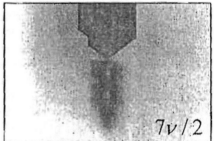

Fig. 9.8 Right: USB-image of 5 μm-wide fatigue crack in Ni-base super-alloy; left: crack photo. After Solodov [9]

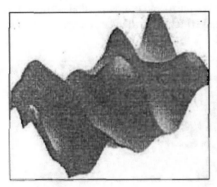

Fig. 9.9 Nonlinear imaging of impact damage in central part of GFRP plate; left-linear (20 kHz-Image); right-UFP-image. After Solodov [9]

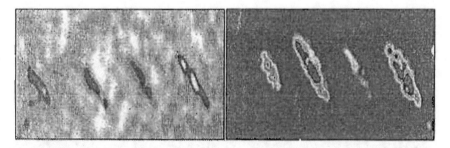

Fig. 9.10 Nonlinear imaging of an impact induced damage in multiply (+45°; −45°) GFR-plate; NLV (left); NACE (9th–11th) higher harmonic image (right) After Solodov [9]

Figure 9.9 left is the linear image at a driving frequency of 20 kHz. It reveals only a standing wave pattern over the whole sample. Figure 9.9 right is the image at the first UFP side-lob of the 10th harmonic of the driving frequency of 198.8 kHz. It shows a clear indication of the damaged area.

The variation of optical reflectivity is a weakness of the scanning laser vibrometry. An example in the damage area with strong scattering of laser light fails in the measurement. Solodov and Bussey [7]'s experiments demonstrated localized sources of nonlinear vibration are planar effects which effectively radiate the nonlinear airborne ultrasound. They [7] propose this nonlinear air-coupled emission (NACE) as an alternative methodology for the location and the visualization of the defects in NDE. In many cases, NACE shows superiority.

A HF focused air-coupled (AC) ultrasonic transducer is used as a receiver for a practical version of the NACE for nonlinear imaging of defects [8].

Figure 9.10 shows the NACE imaging results compared with the NLV of multiple impact damage on the reverse side of a carbon fibre reinforced (CFR) multiply (+45°; −45°) composite plate of 175 × 100 × 1 mm. Similar sensitivity is shown by both techniques which reliably visualize the defects.

The 9th to 11th harmonic NACE image of a 50 μm-wide fatigue crack in a steel plate of 150 × 75 × 5 mm with two horizontally located grip holes for cyclic loading at some distance from the crack is shown in Fig. 9.11. This image reveals the meaning of NACE defects which represent not only the crack itself but also include the fatigue structural damage in the plasticity areas between grip hole and the crack.

The sensitivity of NACE to micro-damage induced by plastic deformation is verified by implementing NACE inspection to a steel specimen with a cold-work area of 5 × 40 mm produced by hammer peening. The imaged at the right of Fig. 9.11 confirms that NACE clearly discerns the micro-damage induced by plastic deformation and it develops even without seriously cracked defects.

Fig. 9.11 NACE imaging in steel specimens: (9–11th) harmonic imaging of 50 μm-wide fatique crack (left); (5 × 40 mm) hammer peening area in steel plate (right). After Solodov [9]

9.5   Conclusions

The feature of localized CAN enables 2D acoustical imaging of nonlinear excitations confined inside the defect areas. Nonlinear NDT (NNDT) of imperfect materials via CAN is inherently defect-selective. That is, it distinctively responds to fractured flaws. Fortunately, this group of flaws includes the most malignant defects for material strength: delaminations, impact and fatigue damages, debondings, icro-and macro-cracks.

Numerous case studies prove the applicability of CAN for defect-selective imaging and nonlinear nondestructive evaluations (NNDE) in various materials by using NACE and scanning NLY. Successful examples in particular include constructional and hi-tech materials such as delamination in fibre-reinforced metal laminates and concrete, delamination in fibre-reinforced plastics, fatigue cold work and micro-cracking in metals.

References

1. Solodov, I. 1998. Ultrasonics of nonlinear contacts: Propagation, reflection and NDE applications. *Ultrasonics* 36: 383–390.
2. Pecorary, C., and I. Solodov. 2006. Non-classical nonlinear dynamics of solid interfaces in partial contact for NDE applications. In *Universality of Non-classical Nonlinearity with Application to NDE and Ultrasonics*, ed. P. Delsanto, 307–3214. New York: Springer.
3. Johnson, P., and A. Sutin. 2005. Slow dynamic and anomalous nonlinear fast dynamics in diverse solids. *Journal of the Acoustical Society of America* 117: 134–140.
4. Solodov, I., J. Wackerl, K. Pffleiderer, and G. Busse. 2004. Nonlinear self-modulation and subharmonic acoustic spectroscopy for damage detection and location. *Applied Physics Letters* 84: 5386–5388.
5. Solodov, I., and B. Korshak. 2002. Instability, chaos, and "memory" in acoustic wave-crack interaction. *Physical Review Letters* 88 (014303): 1–3.

6. Solodov, I., K. Pfleiderer, and G. Busse. 2006. Nonlinear acoustic NDE: Inherent potential of complete nonclassical spectra. In *Universality of Non-classical Nonlinearity with Application to NDE and Ultrasonics*, ed. P. Delsanto, 465–484. New York: Springer.
7. Solodov, I., and G. Busse. 2007. Nonlinear air-coupled mission: The signature to reveal and image micro-damage in solid materials. Applied Physics Letters 91: 251910.
8. Solodov, I., and G. Busse. 2008. Listening for nonlinear dw efects: A new methodology for nonlinear NDE. In *Nonlinear Acoustics-Fundamentals and Applications*, ed. B.O. Enflo, et al., 569–573. College Park: AIP.
9. Solodov, I. 2009. Nonlinear acoustic NDT: Approaches, Methods and Applications [Preprint].

Chapter 10
Modulation Method of Nonlinear Acoustical Imaging

10.1 Introduction

Nonlinear acoustical imaging is based on the physics of nonlinear acoustics. It uses techniques from nonlinear acoustics and is an extension of acoustical imaging from linear to nonlinear regime. It is concerned with the application of finite-amplitude sound wave as well as the nonlinear property of the material or the medium of propagation.

Based on the various techniques from nonlinear acoustics, there are several forms of nonlinear acoustical imaging as follows:

1. Acoustical fractal imaging with applications in medical ultrasound and nondestructive evaluation.
2. B/A nonlinear parameter acoustical imaging with application in medical ultrasound.
3. Harmonics imaging with applications in medical ultrasound, nondestructive evaluation and underwater acoustics.
4. Nonclassical nonlinear acoustical imaging: subharmonics, with application to nondestructive evaluation.
5. Modulation method of nonlinear acoustical imaging with application to nondestructive evaluation.

In this chapter, only the modulation method of nonlinear acoustical imaging will be dealt with.

10.2 Principles of Modulation Acoustic Method

There have been increasing activities in nonlinear acoustical imaging to nondestructive evaluation and medical ultrasound since the early 1990s. Here the modulation acoustic method [1] with application to nondestructive evaluation will be described.

© Springer Nature Singapore Pte Ltd. 2021
W. S. Gan, *Nonlinear Acoustical Imaging*,
https://doi.org/10.1007/978-981-16-7015-2_10

It is applied to the location of a crack in a testing sample. A high acoustic nonlinearity is exhibited by crack [2, 3] which has highly nonlinear properties. Various acoustic responses are produced by cracks such as the generation of higher harmonics and the nonlinear modulation of an ultrasound wave passing through or reflected from a crack by the low frequency vibration tested objects.

Sofar the modulation acoustic method has been applied only to nondestructive testing. Simple nonlinear acoustic technique can be applied only to detect damaged objects. They are unable to provide information on the crack location. The modulation acoustic method on the other hand is capable to locate the crack position in a testing sample. Nonlinear acoustic parameters of solids are much more sensitive to crack-like defects than the linear acoustic parameter such as sound velocity. Differential nonlinear acoustic responses such as higher harmonics generation or frequency mixing can be produced when ultrasonic waves of different frequencies pass through or reflected from a crack. The modulation acoustic method is based on the effect of the modulation of a high frequency ultrasound wave passing through a crack by the low frequency vibration of the test sample. When propagating through the crack, the parameters of the high frequency probe wave are changed by the low-frequency sound wave. Informations on the interaction of high and low frequency sound waves are contained in the modulation coefficient or modulation index. This modulation index is dependent on the position of the crack relative to the nodes and antinodes of the excited low frequency vibration modes in a test sample.

10.3   The Modulation Mode of Method of Crack Location

The modulation method for crack location [1] is based on the result of the interaction of the high frequency ultrasound probe wave and the low frequency vibration of a test sample at the crack. The outcome is the nonlinear force generating the modulation frequency component acting at the position of the crack. A set of modulation amplitude responses from the test sample can be obtained by the excitation of different low frequency modes in the object. Since the resonance properties of the high frequency ultrasonic waves are usually not as strong as the low frequency vibration, the influence of those resonances on the modulation indices can be easily avoided by the use of additional local spatial or frequency averages. Hence one can reconstruct the crack position by measuring the modulation indices for different low frequency modes.

The following is an illustration for the one-dimensional case. For example, take a rod with free boundary conditions at both ends, the low frequency longitudinal modes in such a sample can be given as:

$$u_n(z, t) = A_n \sin \frac{\pi}{l} nz \, \cos \Omega_n t, n = 1, 2, 3 \ldots \tag{10.1}$$

where $u_n(z, t)$ = displacement in the sample, l = sample length, A_n = amplitude, and Ω_n = resonance frequency of the modes.

Let u_ω represent the displacement in the rod caused by the high frequency wave propagating in t. The low frequency and the high frequency waves do not interact with each other if there is no crack in the rod. Interaction takes place if there is a crack in the sample. A modulation effect will be produced which is proportional to the product of $u_n(z, t)$ and u_ω, subjected to the approximation of quadratic nonlinearity:

$$\widetilde{u}_n = \alpha . u_n(z_0) u_\omega \tag{10.2}$$

where $\widetilde{u}_n$ = amplitude of the wave generated by the nonlinear force in the crack at the combination frequency components (modulation frequency components) $\omega \pm \Omega_n$. If one assumes that the newly generated high frequency waves at the combination frequencies propagate in the medium as in an infinite one D medium, then the modulation index is defined as

$$M_n = \frac{1}{A_n} /\widetilde{u}_n |u_\omega/ = \frac{1}{A_n} /[\alpha . u_n(z_0) u_\omega / u_\omega]/ = \alpha[/u_n(z_0)/]/A_n \tag{10.3}$$

Then the parameter is introduced as

$$M(z, z_0) = \sum_n /M_n \sin k_\Omega z/ = \sum_n /\frac{\alpha/u_n(z_0)/}{A_n} \sin k_\Omega z = \alpha \sum_n / \sin k_\Omega z_0. \sin k_\Omega z/ \tag{10.4}$$

where $k_\Omega = \pi n/l$.

Equation (10.4) shows that the parameter M has a peak value at the position of the crack, at $z = z_0$. A second peak takes place at $z = l - z_0$. The spatial resolution of the modulation method depends on the number of excited modes of the flexural oscillation.

10.4 Experimental Procedure of the Modulation Method for NDT

An example of the application of the modulation acoustic method is to nondestructive testing such as the diagnostics of concrete beams. Figure 10.1 shows a scheme of the experiments with concrete neam. Here four concrete beams were used. These four beams have conditions as follows: Beam 1 with a spherical flaw of 3 cm diameter; beam 2 with a transverse artificially made crack; beam 3 with no defects and is used as a reference; beam 4 has an inner reinforcement. The modulation of high frequency 16 kHz acoustic waves generated with a piezoelectric transducer by a low frequency flexural beam vibration excited with the vibrator at the resonance frequencies of the first and the second modes of beam. was studied [1]. Figure 10.2 shows the corresponding diagram of beam oscillation.

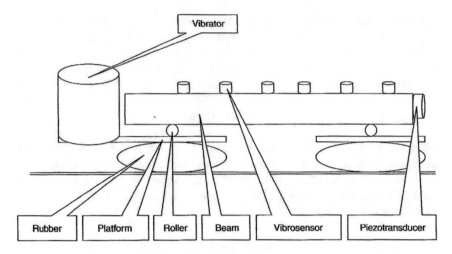

Fig. 10.1 A scheme of experiments with concrete beams (After Didenkulov et al. [1])

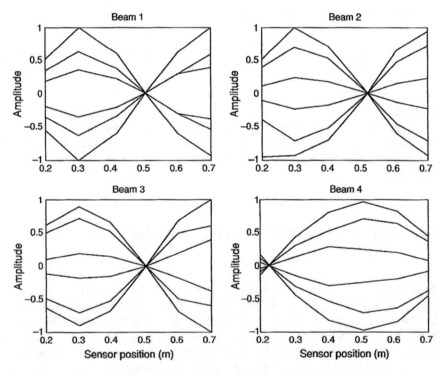

Fig. 10.2 Diagrams of flexural resonance vibrations of beams (After Didenkulov et al. [1])

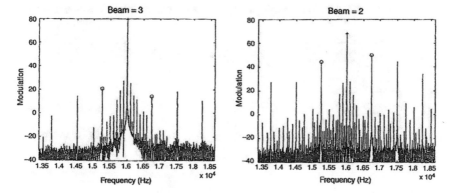

Fig. 10.3 Spectra of signals registered by sensors on the reference beam (left) and on the beam with crack (After Didenkulov et al. [1])

The defects in the concrete beams generate the nonlinear interaction of high frequency and low frequency acoustic waves producing a modulation effect. Figure 10.3 shows the modulation effect revealed by a raising of the lateral frequencies in the spectra of signals from sensors. It can be seen from Fig. 10.3 that the modulation effect is complex. The nonquadratic nonlinearity of the crack produces many lateral frequency components. The difference between levels of the first lateral modulation frequency components and the high frequency components is given by the modulation index. The modulation index can be used as a criterion for nondestructive testing as is shown by measurements done for all the concrete beams.

The experiment shows that the modulation index to depend on the position of a sensor along the testing beam. It is also correlated with the distribution of low frequency modal oscillation along the beam. However, this simple example of the modulation method is not able to detect the positions of cracks in a damaged sample.

10.5 Experimental Procedures for the Modulation Mode System

The modulation mode system has to be used in order to detect the crack position in a damaged sample. This is illustrated by an experiment with a metal rod. Figure 10.4 shows the experimental setup.

The duraluminium rod with 2.1 m length was fastened by two ropes to the support. There was a crack at 50 cm from one end of the rod. A high frequency piezoceramic transducer was glued to one end of the rod with a sensor at the other end. This experiment set provided free end-boundary condition to the rod. Continuous longitudinal sound waves at a frequency of about 200 kHz was emitted by the transducer into the rod. The shock of a hammer excited longitudinal resonance mode in the rod. They were generated simultaneously. All the modulation indices can be measured

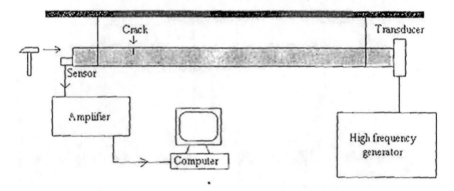

Fig. 10.4   A scheme of the experiment with a metal rod (After Didenkulov et al. [1])

in a single experiment with such a technique because the spectrum of the registered signal contains modulation frequency components for all the modes. In this experiment, the rod has free boundary conditions at both ends. The same function as in Eq. (10.1) can be used to describe the longitudinal modes in such a sample.

Hence a modified technique is used to reconstruct the crack position in a rod. One introduces a modified parameter $\tilde{M}$ instead of using the parameter M. $\tilde{M}$ is given as

$$\tilde{M}(z, z_0) = \sum_n M_n \sin k_\Omega z = \sum_n / \sin k_\Omega z_0 / \sin k_\Omega z \qquad (10.5)$$

It can be seen that $\tilde{M}$ is not as positive as M. It has two peaks: one at the crack position $z = z_0$ and the other which is imaginary at position $z = l - z_0$. Equation (10.5) is used for the reconstruction of crack positions in the rod and is given in Fig. 10.5.

10.6   Conclusions

The nonlinear acoustic modulation method can be used for the detection of cracks exhibiting nonlinear properties. The mean modulation method index can be used as a criterion for the detection of damaged sample. The modulation mode method has to be used to detect the crack positions. Based on the measurements of the modulation indices for resonance low frequency modes in the sample, this method allows one to reconstruct a crack position.

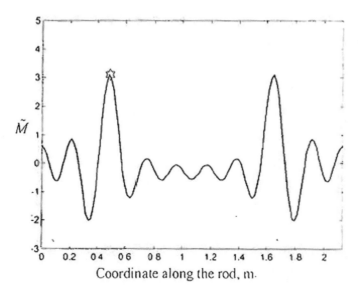

Fig. 10.5 Reconstruction of crack position in the rod. A star marks the real crack position (After Didenkulov et al. [1])

References

1. Didenkulov, I.N., N.V. Kurochkin, A.A. Stromkov, and V.V. Chernov. 2004. Nonlinear acoustic methods for crack vision. In *Acoustical Imaging*, eds. W. Arnold, S. Hirsekorn, 289–296. The Netherlands: Kluwer Academic Publishers.
2. Buck, O., W.L. Morris, and J.N. Richardson. 1978. Acoustic harmonic generation at unbounded interfaces and fatigue cacks. *Applied Physics Letters* 33: 371–372.
3. Sutin, A.M., and V.E. Nazarov. 1995. Nonlinear acoustic methods of crack diagnostics. *Radiophysics and Quantum Electronics* 38: 109–120.

Chapter 11
Applications of Nonlinear Acoustical Imaging and Conclusions

11.1 Introduction

There are a list of forms of nonlinear acoustical imaging such as:

1. Fractal imaging
2. Harmonics imaging
3. B/A nonlinear parameter acoustical imaging
4. Nonclassical nonlinear acoustical imaging
5. Modulation method.

There are three main applications of nonlinear acoustical imaging: nondestrutive testing, medical imaging, and underwater acoustical imaging. For each area of application, there are the particular forms of nonlinear acoustical imaging which are in use. For instance, in nondestructive testing, the harmonics imaging, nonclassical nonlinear acoustical imaging and modulation method are implemented. In medical imaging, the harmonics imaging, fractal imaging and B/A nonlinear parameter acoustical imaging are in use and in underwater acoustics, the harmonics imaging is applicable.

The applications of each form of nonlinear acoustical imaging are mentioned towards the end of each chapter describing the particular form of acoustical imaging. The two particular inventions of nonlinear acoustical imaging which have great potentials for further development in commercial applications are the acoustical fractal imaging and the B/A nonlinear parameter acoustical imaging. They will be used mainly in medical imaging.

© Springer Nature Singapore Pte Ltd. 2021
W. S. Gan, *Nonlinear Acoustical Imaging*,
https://doi.org/10.1007/978-981-16-7015-2_11

Printed in the United States
by Baker & Taylor Publisher Services